GOD IS!

why evolution isn't

A GOD-Centered Christian Response
to the Creation-Evolution Controversy

A. ARTHUR PINNO

Copyright © 2011 by A. Arthur Pinno

God Is!
why evolution isn't
by A. Arthur Pinno

Printed in the United States of America

ISBN 9781612156200

All rights reserved solely by the author. The author guarantees all contents are original and do not infringe upon the legal rights of any other person or work. No part of this book may be reproduced in any form without the permission of the author. The views expressed in this book are not necessarily those of the publisher.

Unless otherwise indicated, Bible quotations are taken from THE HOLY BIBLE, NEW INTERNATIONAL VERSION®, NIV®. Copyright © 1973, 1978, 1984, 2010 by Biblica, Inc.™ Used by permission. All rights reserved worldwide.

www.xulonpress.com

Table of Contents

FOREWORD: DR. FRAN MONSETH xiii
A WORD FROM REVEREND DR. HARALD SCHOUBYE xvii
ACKNOWLEDGEMENTS xix
INTRODUCTION: AN INVITATION TO REASON................ xxi

PART A: SCIENCE, EVOLUTION AND GOD

CHAPTER 1:   A GOOGOLPLEX OF COMPLEXITY
              TESTIFIES TO GOD............................29

    The Unfolding Miracle of What Is................29
    The Cosmos Chorus or the Sky
        Screamo-Band 32
    Cellular Central 39
    DNA Dynamics 40
    One Person's Junk is Another's
        Treasure 42
    Do You Think? 45
    Irreducibly Complex Features in All
        Creatures...46
    Signs of Intelligence.....................................51
    The GOD of Miracles! 55
    Psalm 96 55

CHAPTER 2: GOD VS. NOTHING- Why Evolution Flunks Logic, Math, and Science 57

Origins: GOD or Nothing? 57
Philosophically Lucky Math or an Incredible Creator? 62
Information Isn't Matter, It Is Mind over Matter 63
The Facts of Life vs The Myths of Life 69
Are You Related to the Apes? 72
Mutations: The Engine or the End of Evolution? 76
A Flood of Fossils 80
Where are the Ape-People and Other Mutating Creatures? 82
Miracle Blood or a Mirage of Time? 85
The Population Bomb Explodes Evolutionism 87
Time to Turn the Page 90

CHAPTER 3: SCIENTISTS AREN'T GOD & SCIENCE IS NO SAVIOR 91

Humanism and the Worship of Science 92
Children in the Little Leagues 95
Creation Scientists of the Past 97
Beware the Boasting of the White Coats ... 100
Faith in the Scientific Experts 103
Happy ___Birthday to the Earth 109
When Time Flew and Light Stretched 113
Scientific Confusion on Testable Matters ... 117
The Scientific Record: Gems of Truth Mixed with Fool's Gold 119
What Tornadoes, Explosions, Water and Wind Can't Explain 120

Is Evolution A Massive Worldwide
Conspiracy?..123
When Hubris Leads to Blindness..............127
The Materialistic Tunnel Vision of
Science 128
Science: Helpful, Neutral, Destructive,
or All of the Above?............................130

PART B: THE ONE AND ONLY GOD

CHAPTER 4:   GOD IS 135

What is Infinite?..135
Knowing GOD...136
GOD IS!..138
GOD IS Eternal! 139
GOD IS Almighty! 141
GOD IS Infinite in Intelligence and
Wisdom!..144
GOD IS Holy and Perfect!..........................145
One GOD, Many Idols 148
Can Someone Believe in CHRIST and
Evolution?..149
If GOD Were Moloch, Evolution Could
Be His Modus Operandi 153
What is Wrong with Theistic Evolution?...154
The Living GOD is not a Little Tinkerer!....157
Does the End Justify the Means?..............157
Unbelief in the Scriptures Leads to
Ignorance of GOD 159

CHAPTER 5:   GOD IS LOVE & LIFE 162

PART A: The GOD of Love and Life or the Little
Tinkerer of Misery and Death?...........162

The Love of GOD 162
Little Tinkerer: A Supreme Mad Scientist
    and Gore Addict 167
JESUS is not a Little Tinkerer 168
Is Death Physical or Spiritual in
    Genesis 2- 3? 170
What About Lassie and Flipper? 172
Roses are Red and Cherries are
    Delicious 174
Mirror, Mirror on the Wall, Who is the
    Culprit of It All? 175
Why Does GOD Tolerate Evil in our
    World? 179
The Limits of GOD'S Tolerance 183
Alternatives to Tolerance 184

PART B: Can GOD'S Love and Hell Co-exist? 186
Could a GOD of Love Send Anyone to
    Hell? 186
What is Hell? 187
Are Prisons Good? 196
Hallelujah! Heaven is Around the
    Corner 198
The Way Out and Up 201

CHAPTER 6: "GOD IS NOT A MAN THAT HE
    SHOULD LIE" 203

"I Tell You the Truth"- JESUS, the CHRIST ... 203
FAITH is Trusting GOD'S Word 207
JESUS and Adam 210
Should Genesis be Taken Literally? 212
When Is a Day a Day? 213
Was JESUS Deceived and a Deceiver? 216
JESUS Testified to the Flood 217

The Flood of Noah's Day Washes Away
    the Evolutionary Muck 218
The Rainbow: the Sign of GOD'S
    Faithfulness or Deceit? 222
All Truth is GOD'S Truth 224
GOD is the Miracle Working GOD 226
The Evolutionary Scenario Has No Room
    For CHRIST in its Inn 228
Where There is No Sin There is No Need
    for a SAVIOR 229
The Truth is Marching On and Setting
    People Free 232

PART C: FRUITS AND CHOICES

CHAPTER 7: BY THEIR FRUITS 237

JESUS' Words of Warning 237
The Fruits of the Evolutionary Tree 239
Contempt for GOD, His Word and His
    Church 240
Atheism: The Logical Fruit of Evolution ... 244
No Gods, Many Gods, or the One and
    Only GOD? 255
The Bloody Record of the 20th Century 257
From Darwin to Hitler: Connecting
    the Dots 259
"There is a Way that Seems Right to a Man,
    but in the End it Leads to Death" 264
Sacrificing Children to the Evolutionary
    Goddess 265
More to Come 268
The Tree of Life and the Fruits of the
    SPIRIT 269

CONCLUSION: MULTITUDES IN THE VALLEY OF DECISION 271

He's Everything to Me 271
Kairos Time for Choosing GOD 272
GOD and Science 275
The Fence Sitters 276
Different Universes 277
GOD IS! 278
This is My GOD- Heidi Pinno 279
Good News from Our Creator 283

ENDNOTES 285

BIBLIOGRAPHY 293

**Glory to GOD in the Highest
And on earth
Peace to all People
On Whom His Grace Rests!**

FOREWORD

The book you are holding needed to be written. With the continual bombardment of propaganda promoting the theory of evolution, many Christians have become confused and have begun to wonder if the biblical account of the earth's origins should be taken in a literal or in a symbolic sense. Sadly, some have conceded to the prevailing pressure to surrender their prior belief that the first chapters of Genesis should be understood as straightforward historical narrative. While continuing to believe that God initiated the process at some point in the distant past, they have succumbed to the alleged "facts" of science regarding the longevity of earth's origins.

It is no wonder that someone who does not take the Bible seriously is typically an advocate for Charles Darwin's "old-earth" theory. For the secular mind-set, an alternative simply must be found for the God-centered account of creation in the opening pages of Scripture. There are too many unsettling implications for the unbeliever if credibility should actually be given to the biblical text, implications that may actually lead to a radical change of mind regarding existence on this little planet and one's purpose for being here.

The author of this timely volume ably and amply illustrates that, for a Christian who believes the entire Bible is

God's inspired and inerrant Word, there can be no worthy alternative explanation for the earth's origin than what is presented in Genesis and referenced throughout both Testaments. It is the accuracy as well as the authority of Scripture that is at issue in this matter. The author offers an abundance of biblical citations which demonstrate clearly how the Bible is literally replete with statements exalting the wisdom and power of God in creation, unfolded in sober and simple narrative in Genesis.

The author is a Bible-believing pastor. He accepts by faith the internal testimony of Scripture regarding its full inspiration and inerrancy and is heartened by the external evidence supporting this testimony. His earnest endeavor is to exposit the Scriptures he knows and loves with a special focus on God's mighty and majestic creation. On page after page, the reader will find biblical references with concise commentary and application.

Although the author's purpose is not to present a scientific treatise in seeking to uphold the biblical account of creation, he demonstrates a thoroughgoing familiarity and understanding of contemporary thought relative to the evolutionary theory. With a gentle forcefulness, he exposes inconsistencies and "leaps of faith" taken by evolutionists in their anxiety to promote their anti-theistic and unbiblical theory.

While some in the Christian community today seem to treat the doctrine of creation as ambiguous and as such, relatively unimportant, the author is forthright and persuasive in underscoring the clarity and the vital importance of this biblical truth, indeed, a fundamental doctrine of the Christian faith! He conclusively shows the inextricable connection of the doctrine of creation with all of the major doctrines of God's Word and emphasizes the devastating consequences of surrendering the Bible's teaching on this critical subject. It is altogether predictable (and sadly, often inevitable) that a believer who begins to treat the Genesis account of creation as anything but historical narrative will at last surrender everything he once believed!

With a deep concern to warn fellow-believers of the danger of the seemingly smallest concession regarding the truth of God's creation, the author shares his Bible-based convictions with clarity and compassion. Ultimately, it is a matter of taking God at His Word, simply believing that what the Holy Spirit has inspired the penmen of Scripture to record is absolute truth in every respect including the doctrine of creation.

Dr. Fran Monseth, Dean, Association of Free Lutheran Congregations Seminary, Plymouth, MN U.S.A.

A WORD FROM REVEREND DR. HARALD SCHOUBYE

A. Arthur Pinno has been thinking about this project for a number of years. He has wanted to alert people of all ages but particularly the young adults to the falsehoods that proponents use to support evolution. The concept of evolution defines the secular age. There are only two alternatives that people have for how things came to be—creation by an eternal personal intelligence or manufactured by self-generation out of nothing. People resort to the second only after a conscious decision to exclude GOD from explanations for the origin of the world. Evolution is the best way for someone committed to naturalism to explain how the world came to be but it is a weak position. The weaknesses of this approach are hidden to the public. The author takes these weaknesses out of hiding. He has endeavored to present the evolutionists fairly and accurately. He hopes that this book will generate discussion, convert the evolutionist, alert the naïve, and strengthen and arm the creationist. His book is necessary reading today and will become more important with time as secularism holds sway. I commend this book to everyone who wants to know the truth about the belief in evolution and creation.

- Dr. Harald Schoubye is the pastor of Living Word Lutheran Church, Vernon, British Columbia, Canada.

ACKNOWLEDGEMENTS

To the eternal GOD: Thank you my LORD and my GOD for the gift of life and the amazing beauty and intricacy of all of nature that You have surrounded us with. Your glory fills the universe and we give You thanks. You are AWESOME! There is no one like You. I know that many charge You with being incompetent and unjust, even cruel, because of Your temporary tolerance of evil. If only Your accusers would humble themselves and come to know You, the infinite I AM, how blessed they would be. Thank You for Your long suffering patience with humanity and with me, the patience which only You could have with us. If You had chosen to put an end to the crimes, foolishness, sins and arrogance of the human race, even a hundred years ago, to have prevented the great evils of the twentieth and now the twenty-first century, I would never have had the opportunity to experience the life You have given me and to have come to know You, the one true GOD, and JESUS CHRIST Whom You sent into our world. Thank You for Your amazing grace, peace, and eternal hope that You give so freely and abundantly to all who trust in You. Thank You for Your eternal Word, through which You have brought all things, visible and invisible, into existence. Your Word is a bright and comforting light in the darkness of our world. Thank You for the blessings of family. Thank you for parents, whose love and faithfulness to You, to each other, and to their family taught me a lot about who You are. Their songs and faith continue in my heart. Thank you for the gift of my wonderful wife; Your beautiful daughter You've given to be my partner, companion, and mother of our children. Thank You for the blessing of our children who, like us, need Your grace and help for each day as they too run the race (sometimes walking, sometimes crawling, and sometimes just holding on) through the valleys, deserts, plains, meadows and mountains of life. What a joy they are to us! Thank You for filling our lives with Your grace.

Thank you to all of the faithful men and women who have persistently and unashamedly proclaimed the foundational truth of creation as revealed in GOD'S Word. Specifically I want to thank Creation Ministries International, Answers in Genesis, and the Institute for Creation Research. Despite the ridicule and contempt by the scientific establishment, their unflinching faith in GOD'S Word and the joy of discovery of new insights into GOD'S awesome creation have been a tremendous encouragement. I have also been blessed by the Intelligent Design network of scientists and philosophers and their courage to challenge evolutionary orthodoxy in the scientific community, often at the price of ostracism and verbal and reputational abuse. May they come to know the infinite Creator behind the amazing designs.

Thank you to Dr. Fran Monseth and Dr. Harald Schoubye and my wife, Dorothy, for their assistance in editing these pages and helping me to keep the thoughts clear and concise.

LORD GOD, my Creator and Savior, may You bless these loaves and fish for Your glory and for the strengthening of Your people and the salvation of all who read it! May it be so!

- A. Arthur Pinno, Epiphany January 6, In the Year of our LORD 2011

INTRODUCTION

An Invitation To Reason

"Come now, let us reason together," says the LORD.
(Isaiah 1:18)

Most people are not reasonable, or, at least, most people think that most other people are not reasonable. What is considered to be reasonable is usually dependent on an individual's perspective of what is real. This is certainly true regarding the creation-evolution issue. For those who believe that there is no GOD, it is not reasonable to ponder the possibility of GOD as the omnipotent creator of all things. When GOD is rejected as the Creator, evolution is upheld as the only reasonable option. On the other hand, there are many who do believe that there is a GOD who had everything to do with the creation of our universe and world. For those who believe in GOD, as I do, it is difficult to comprehend how anyone could be so blind and unreasonable as not to recognize the hand of GOD in our amazingly complex universe.

Reason is the foundation of the scientific endeavor. The ability to reason is necessary to make deductions from observations and to take ideas or premises to their logical conclusions. It is absolutely essential for living and for doing science. Without a mind and spirit it would be impossible to reason. There would be no you or me to do any reasoning if we did not have a mind and spirit. Since thought is not a material reality, material atoms alone are incapable of reasoning. Reason is a quality of the mind and spirit, not of material. Anytime secular materialists take part in a reasoned debate on the reality of GOD, they lose just by showing up. If human life was only a conglomeration of chemicals, then debates would be nothing more than atoms blowing off gas. Descartes' famous dictum *"Cogito, ergo sum"* – "I think, therefore I am" expresses the reality that without thought there would be no consciousness of reality or identity. Our ability to reason is a testimony to the reality of our mind and spirit, which in turn are testimonies to the Supreme mind and Spirit that we owe our existence to. It is GOD who enables, invites and encourages us to reason and to seek after truth and the way to the true, abundant, and eternal life.

Is material evolution or is the Genesis creation account a reasonable explanation for the marvelous realities that surround and encompass us? Is theistic evolution the answer that harmonizes science and religion? What are the presuppositions and premises of atheistic or theistic evolution and what are the logical conclusions that they lead to regarding life and GOD? What are the implications of the reality of the infinite nature and character of GOD for our understanding of creation? How does JESUS, the center of Christian faith, fit in with the creation-evolution issue? Could JESUS be the mind behind evolution? This book is all about these and other related questions that invite and engage our ability to think and to seek out a reasonable and consistent Christian understanding of the origin of our universe and our lives.

Like many in my generation and the generations that have followed, I was strongly influenced by our secular public education system and the secular media into accepting standard evolutionary theory, with GOD quickly tacked on at the end for traditional ornamental purposes. During my high school and early university years, I temporarily harmonized in my own mind the evolutionary theory with the truth of GOD as Creator. Of course, just because something may fit into one's mind and imagination does not mean that it fits reality. The human mind is capable of incredible imagination with no grounding in truth. I do understand the pressure for such harmonization on those who feel they are caught between contemporary "scientific" theories and the creation teachings of the Christian faith. Fortunately, my education didn't end with high school and university. I continued to question and rethink what I believed and why.

As we go about our lives we are surrounded by a universe and world that point us to GOD. In the past it was the manifest reality of the birds, the bees, the flowers, the trees and all that filled our senses that directed our thoughts to the Supreme Mind who fashioned it all. Today, to these obvious testimonies concerning GOD, we can add these once-concealed truths: the incredibly exact balance between the strong and weak nuclear forces that knit the cosmos together, the multilayered error-checking digital code of the cell, and the almost infinite complexity and unimaginable precision of every part of the created order. Every detail of our cosmos directs our attention to the living and active presence of the infinite GOD. The more I learn about our world and the nature of GOD, the more I realize how empty and shallow the evolutionary myth is and how deceptively it has been leading many away from their Creator and Savior.

These pages are intended primarily for those who are Christians in the historical and Biblical sense. To those who believe and trust in JESUS CHRIST as GOD'S SON and the only SAVIOR and LORD, but have been subjected to the profuse propaganda of evolution through secular media and educational systems, I hope that the truths covered here will spur you on in your faith. My prayer is that the LORD would bless this book to encourage you to trust in Him and His Word and to follow Him on the path that leads to the abundant and eternal life.

Should those of you who are convinced evolutionists read this, I hope that this will encourage you to pause and take a second look, a closer and more skeptical look, at the so-called "facts of evolution". Even more importantly, I hope that you will seek out the GOD who created you, loves you and desires to save you from the hopelessness of the evolutionary mythology and the despair of our fallen world.

Before proceeding further, it is crucial for the reader to understand that the term "evolution" is being used throughout these pages in the fairly broad sense in which it is most often used by the general public. Throughout this book "evolution" is used in reference to the whole material process through which some believe that all matter, and even life itself, came into existence and changed and developed. This includes both cosmic and biological evolutionary beliefs. In regards to life, particularly, I am not referring to minor changes in species as a result of genetic variation, but to the Darwinian tree of life: the belief that all life forms are descended from a single cell through the transformation from one kind of creature to another by natural processes. Evolution is, as some creation scientists have poetically expressed it, "from goo to you via the zoo".

Evolution is a collection of mythological stories meticulously spun together from materialistic assumptions in order to explain the reality of our world and life without any help from GOD. The typical mantra for biological evolution is that through natural selection, mutations and the magical kiss of time over thousands, millions, even billions of years, all of the incredibly complex and amazing creatures that inhabit planet Earth have spontaneously evolved via natural laws, without any supernatural intervention.

Undoubtedly, it is the crowning achievement of secularism to have developed such an elaborate theory and to have misled so many in Christian circles into subjecting their beliefs to the vagaries of evolutionary imagination. Thankfully, in the light of CHRIST, His Word and true reason, many are being set free from the myths of secularism and its evolutionary foundation. **JESUS said: "If you continue in My Word, you will know the truth and the truth will set you free" (John 8:31,32).** It is good to be free from false paradigms and philosophies that reduce and minimize the importance and reality of GOD and of human life created in His image.

I am not a scientist, but a Christian believer called to proclaim the wonderful truth of GOD and CHRIST and His eternal Kingdom. This is not a scientific book attempting to scientifically establish the truths of the Christian faith. I know of no one who believes that the Bible is a scientific book or that it claims to be a scientific book. The Bible is a book of faith: faith in the GOD who created humanity, and who has been working down through history to win and to save all who seek Him and turn to Him, choosing individuals, families, and even nations to enter into a covenant relationship with Him. The true teachings, history, and doctrines of the Christian faith centered in CHRIST are nevertheless in complete harmony with the ultimate scientific facts concerning our life, world and universe. GOD is sovereign over all things. He is the GOD of science and, when all is said and done, His Word will be found to be in perfect alignment with the scientific materialistic realities of life.

In these pages I am sharing some very basic logical, evidential, scientific and Christian Biblical reasons why people shouldn't be deceived into accepting the evolutionary tale as a fact of life. These reasons are not in any way exhaustive, but I believe they do provide more than enough information for rejecting the storyline of evolution.

Evolution is a sand castle or house of cards that will soon implode. The unimaginable complexity of our universe and of life is overwhelming all but the most doctrinaire of evolutionary believers. Those with open minds will soon, if they haven't already, catch a glimpse of the infinite mind of GOD revealed in our amazing world and universe. Hopefully, that will be a springboard to living in the reality of our awesome Creator and Savior. May it be so!

The material presented in this book consists of truths that GOD has taught me over the course of my life. I have sensed GOD'S direction in writing this book and have sought to be faithful to His Word. However, I also acknowledge that my own understanding of all things in heaven and earth is limited and that as a child of GOD I have a lot to learn in every area of life. As His Word says: *"For now we see only a reflection as in a mirror; then we shall see face to face. Now I know in part; then I shall know fully, even as I am fully known." (I Corinthians 13:12).* I believe that if you carefully consider the truths presented here and test them by the revealed truth of GOD'S SPIRIT in His Word, you will be challenged and encouraged. May the HOLY SPIRIT of the eternal GOD give you His wisdom and discernment as you engage your mind in contemplating these truths. To GOD be the glory!

> *"Come now, let us reason together," says the LORD.*
> *Though your sins are like scarlet, they shall be white as snow;*
> *Though they are red as crimson, they shall be like wool." Isaiah 1:18*

PART A

SCIENCE, EVOLUTION AND GOD

Chapter 1

A Googolplex of Complexity Testifies to GOD

"What is impossible with men is possible with GOD."
JESUS Luke 18:27

THE UNFOLDING MIRACLE OF WHAT IS...

Have thought of it, could you?
 "Not I", confessed the wise:
  a hundred billion cells...trillions of interconnections... a mind to contemplate what is and is not...
 a googolplex of incredible creatures with fascinating features...
the web of the spider...the sonar of the bat...
 vroom.. vroom.. goes the outboard motor in the bacterium's tail
  externalizing its miniaturized computerized program ...
DNA multi-layered codes placing human technology into the toys for tots territory...

salmon and monarch go home—to a place they have not known...
the bombardier beetle...a chemist...an explosive expert.. an Olympian marksman...
gravity: planets orbit,
apples fall,
birds sail over us all...
there is light: what it is, no one understands, but it is...
as the first of GOD'S creation it has been from the beginning...
eyes that enable our minds to see the majesty of it all...
none so blind who choose not to see...
none so foolish who think they could do better.
Awesome GOD,
grant us wisdom to perceive the miracle of what is and the deception of what is not.

People often rhetorically ask: "Can you prove that GOD exists?" Usually they are thinking about a scientific proof, that is somehow capturing GOD and putting Him through a series of tests, as if He were a malignant tumor or a soil sample. As many others have noted, if we could do that, He wouldn't be GOD.

There are googolplexes of reasons to believe in GOD and to reject the evolutionary story. A googol is 10 to the 100th power, a number larger than the number of known atoms in the universe. A googolplex is 10 to the $googol^{th}$ power, an incomprehensible figure, a mathematical representation of infinity. Our incredible world and universe is literally filled with a googolplex of realities that defy evolution and testify to the eternal GOD of creation.

We live in the most fascinating and unimaginable universe and planet that defy human comprehension. Our ancestors could never, in their wildest dreams, have conceived of the spectacular reality in which we live. Neither can we, even as we begin to catch a glimpse of the majesty of what is. There is so much to learn and everything we learn increases the awe. There are not just the Seven Wonders of the World but a googolplex of miraculous features and creatures. Our Father in Heaven has immersed us in a myriad of incredible interrelated intricacies, from the cosmos to the atom, to keep us on our toes, and on our knees. I will boast of what my FATHER in Heaven has done.

In this chapter we will be looking at just a few examples, from the googolplex of possibilities, that clearly testify to the mind of GOD and that defy the random meaningless processes of evolution. The German astronomer, Johannes Kepler, is credited with saying that what he was doing in science was "thinking God's thoughts after Him." In today's technology, one of the areas which is blooming is biomimicry, the imitation of life; imitating GOD'S thoughts. Velcro was inspired by the nature of burrs, passive cooling systems are based on the architecture of termite mounds, wind turbines mimic flippers on humpback whales, bionic cars come in the shape of odd tropical fish for energy conservation, and friction-reducing coatings for the hulls of ships, submarines, and aircraft are based on shark skin. These are just the tip of the iceberg of the designs being discovered in life forms, and then being copied and used for the advancement of human technology.[1] The best way of learning is to look at what GOD has done and try to mimic it as closely as possible.

In sharing just a few examples of the wonders of our cosmos and world, I do not in any way claim to fully understand the scientific complexity involved in these illustrations. No one does, except, of course, GOD. Despite all the boasting on the part of some in the scientific community, life and our universe are still great mysteries. If anything, the mystery is increasing as we learn more.

Like the people of JESUS' day, including the Pharisees and King Herod, many individuals today say, "show me a miracle and then I will believe and worship". What is a miracle? A miracle is something which is outside the realms of human ability and beyond the presently known laws of the universe; something which is impossible for man, but possible for GOD. Our universe, world and everything in it are absolutely miraculous. The universe itself is beyond the presently known laws in the universe. It has been conceived and materialized by a mind and power and will beyond our comprehension. The zeal of the LORD Almighty has accomplished it. Humility and worship are good places for us to start. Here are just a few miracles to get us started on the right path.

THE COSMOS CHORUS OR THE SKY SCREAMO-BAND

"The heavens declare the glory of GOD; the skies proclaim the work of His hands. Day after day they pour forth speech; night after night they display knowledge. There is no speech or language where their voice is not heard. Their voice goes out to all the earth, their words to the ends of the world." (Psalm 19:1-4)

From the beginning of time the heavens - the universe - the cosmos - the sky - the firmament - the expanse - has been a miraculous sign testifying to the amazing power and intelligence of GOD. While many have been deceived into worshiping false gods of the cosmos, like the sun or planets, many others, in every part of the world, have lifted their eyes to the heavens and called out to the living GOD who made it all.

Today, the cosmos is not only singing the glory of GOD but is screaming it out for everyone, especially the hard of hearing. Nevertheless, there has been a strong current flowing through secular science into mainstream culture that believes, and doesn't hesitate to proclaim, that new scientific theories on the cosmos invalidate the ancient belief that GOD created the universe. Nothing could be further from the truth. While there are those who seek to force the new theories regarding the formation of the universe to fit into a strictly materialistic mold, their theories and mold fall far short of the realities which the expanse reveals.

The evidence from the heavens is causing many unbelievers today, even in the scientific community, to shake their heads in bewilderment at the amazing mathematical precision the universe is communicating to us. An article called: "The universe is finely tuned for life," on the Creation Ministries International website, lists a few of the physical realities of our cosmos and world that need to be finely tuned for life to exist on our planet:

> Strong evidence for a Designer comes from the fine-tuning of the universal constants and the solar system, e.g.
>
> - The electromagnetic coupling constant binds electrons to protons in atoms. If it was smaller, fewer electrons could be held. If it was larger, electrons would be held too tightly to bond with other atoms.
> - Ratio of electron to proton mass (1:1836). Again, if this was larger or smaller, molecules could not form.
> - Carbon and oxygen nuclei have finely tuned energy levels.
> - Electromagnetic and gravitational forces are finely tuned, so the right kind of star can be stable.
> - Our sun is the right colour. If it was redder or bluer, photosynthetic response would be weaker.
> - Our sun is also the right mass. If it was larger, its brightness would change too quickly and there would

be too much high energy radiation. If it was smaller, the range of planetary distances able to support life would be too narrow; the right distance would be so close to the star that tidal forces would disrupt the planet's rotational period. UV radiation would also be inadequate for photosynthesis.
- ➢ The earth's distance from the sun is crucial for a stable water cycle. Too far away, and most water would freeze; too close and most water would boil.
- ➢ The earth's gravity, axial tilt, rotation period, magnetic field, crust thickness, oxygen/nitrogen ratio, carbon dioxide, water vapour and ozone levels are just right.[2]

Another name for the fine-tuning of our universe is the Anthropic Principle. Dr. Grant Jeffrey, in his book, *Creation: Remarkable Evidence of GOD's Design,* explains the history of the Anthropic Principle:

> In 1973, a very important scientific conference was held in Poland that celebrated the five hundredth year since the birthday of Nicolaus Copernicus, the first and greatest astronomer of his age. A respected astrophysicist from Cambridge University, Dr. Brandon Carter, delivered a paper called "large Number Coincidences and the Anthropic Principle in Cosmology." Dr. Carter coined the phrase "anthropic principle", derived from the Greek word anthropos, which means "man". Dr. Carter proposed an extraordinary theory: that the only rational way to explain the fact that the universe existed as it does, with an incredibly precise balance between all of the multitude of forces including gravity, electromagnetism, and the strong nuclear force that made our Universe possible, can only be explained if they were fine-

tuned in such a precise manner to allow human life to exist on Earth. [3]

Dr. Henry Morris comments on a few of the fine-tuned realities of our world:

> The earth, with its unique hydrosphere, atmosphere, and lithosphere is, so far as all the actual evidence goes, the only body in the universe capable of sustaining higher forms of life such as man. This, of course, is exactly as would be predicted from the creation model. The earth was created specifically to serve as man's home.[4]

What a magnificent and marvellous home GOD created for us! That is our FATHER for you!

For a more detailed understanding of the fine-tuning of the universe, or the Anthropic Principle, just search it out on the net and you will have plenty to awe you.

Like many others, I certainly can't comprehend all of these precisely tuned realities of our universe and world, but you don't have to be a rocket scientist or an astrophysicist to realize that our planet and universe are miraculous and still very mysterious, with many new discoveries before us. Every dynamic aspect of the universe presents us with a glimpse of the infinite power and intelligence of GOD.

One of the most well known hymns in the Christian Church today was written by Swedish Pastor Carl Boberg in 1886. *"How Great Thou Art"* expresses praise to GOD for the marvels of the universe.

> *O Lord my God! When I in awesome wonder,*
> *consider all the worlds Thy hands have made.*
> *I see the stars; I hear the rolling thunder,*
> *Thy power throughout the universe displayed.*
>
> *Then sings my soul, my Saviour God, to Thee;*

> *How great Thou art, how great Thou art!*
> *Then sings my soul, My Saviour God, to Thee:*
> *How great Thou art, how great Thou art!*

It is not just hymn writers and believing Christians who recognize that our universe is an awesome, miraculous place. Here are some testimonies of various physicists, astronomers, cosmologists and other scientists, most of whom are not Christian believers, but who can't help but marvel at the melody and lyrics the choir of the cosmos is singing today:

Fred Hoyle, astrophysicist: "A common sense interpretation of the facts suggests that a super intellect has monkeyed with physics, as well as with chemistry and biology, and that there are no blind forces worth speaking about in nature. The numbers one calculates from the facts seem to me so overwhelming as to put this conclusion almost beyond question."[5]

George Ellis, astrophysicist: "Amazing fine tuning occurs in the laws that make this [complexity] possible. Realization of the complexity of what is accomplished makes it very difficult not to use the word 'miraculous' without taking a stand as to the ontological status of the word."[6]

John O'Keefe, astronomer at NASA: "We are, by astronomical standards, a pampered, cosseted, cherished group of creatures... If the Universe had not been made with the most exacting precision we could never have come into existence. It is my view that these circumstances indicate the universe was created for man to live in."[7]

Arthur Eddington, astrophysicist: "The idea of a universal mind or Logos would be, I think, a fairly plausible inference from the present state of scientific theory."[8]

Robert Jastrow, astronomer and self-proclaimed agnostic: "For the scientist who has lived by his faith in the power of reason, the story ends like a bad dream. He has scaled the mountains of ignorance; he is about to conquer the highest peak; as he pulls himself over the final rock, he is greeted by a band of theologians who have been sitting there for centuries."[9]

Wernher von Braun, Pioneer rocket engineer: "I find it as difficult to understand a scientist who does not acknowledge the presence of a superior rationality behind the existence of the universe as it is to comprehend a theologian who would deny the advances of science."[10]

As an aside, Dr. Wernher von Braun, who was the lead scientist with NASA in its early years, said, "For my confirmation, I didn't get a watch and my first pair of long pants, like most Lutheran boys. I got a telescope. My mother thought it would make the best gift."[11] The telescope can be a great tool for thinking GOD'S thoughts after Him. As for me, I got a watch, but time is also a great gift from GOD that enables us to contemplate a little of His majesty declared in the heavens above.

Lastly, Dr. Gerald Schroeder, a nuclear physicist, shares this quote from Dr. Michael Turner, an astrophysicist at the University of Chicago and Fermilab, who describes the fine-tuning of the cosmos in this way, "The precision is as if one could throw a dart across the entire universe and hit a bull's-eye one millimeter in diameter on the other side."[12]

These testimonies are from those who see and acknowledge that the mathematical precision of the universe and world we live in points to GOD, but perhaps the loudest testimony to how clearly the Sky-Screamo band is proclaiming GOD'S praise is given by those who refuse to join the heavenly chorus. "Methinks they doth protest too much". Those who insist that the fine-tuning of the universe is mere chance have developed, and are actively promoting, a new theory called the "multiverse" to rationalize GOD out of the picture.

The multiverse is an imaginative theory held by an increasing number of atheistic cosmologists. They believe, with an a priori prejudice against GOD, that the only scientific explanation to account for our finely tuned universe is the existence of an infinite number of universes in other dimensions beyond our universe. Believe it or not, their theory is that there must be some universe-creating force out there, beyond our time-space reality, that is continually spitting out new universes with all kinds of mathematical equations. They believe that, as a result of the infinite number of universes this force has created, we just happen to have gotten lucky to live in a universe that is as perfect as ours. The probability of an ordered universe and life is so infinitesimal that they have now resorted to imagining an infinite number of universes to explain why ours is so precise. When those who reject GOD have to go to the extent of imagining infinite universes, all of which just happen, you know that the Sky-Screamo band of the cosmos is starting to disturb their sleep.

From a Christian perspective, it is certainly possible that GOD may have created other universes beyond ours. We certainly believe in the reality of a new heaven and new earth that GOD will create for His people (John 14:1-6, Revelation 21:1-8). Nevertheless, this is the only universe that human science can know anything about, and it is so finely tuned that the only reasonable explanation for it is an intelligent designer. By conjuring up all kinds of mythical material realities outside of our cosmos, these scientists shut their eyes, cover their ears, and close their minds to the clear cosmic testimony regarding the mind of GOD. While their eyes and ears may be closed, the universe continues to declare the glory of GOD in harmony of screamo dimensions. Their "anything but GOD" protestations and wild imaginations only amplify the volume of the heavenly band.

It is right and good for us to join the heavenly chorus in praise to GOD, saying, singing, and shouting, **"You are worthy, our LORD and GOD, to receive glory and honor and power, for You created all things, and by Your will they were created and have their being." (Revelation 4:11)**

CELLULAR CENTRAL

From the vastness of the cosmos, we redirect our eyes and minds to focus on the microscopic miracles that surround and embody us. Prior to our generation, no human being could have even remotely imagined the miniature world of the cell. As we enter the microscopic realm of life, we find ourselves in an even more magnificent universe than the one above us. The cell is a marvel of miniature design on a level far exceeding our present human capabilities. While the cosmos chorus declares the glory of GOD, the cell magnifies His praise sevenfold.

Molecular biologist, Dr. Michael Denton, who is a non-creationist, non-Christian skeptic of Darwinian evolution, wrote in his book *Evolution: A Theory in Crisis:*

> "Perhaps in no other area of modern biology is the challenge posed by the extreme complexity and ingenuity of biological adaptations more apparent than in the fascinating new molecular world of the cell...To grasp the reality of life as it has been revealed by molecular biology, we must magnify a cell a thousand million times until it is twenty

kilometers in diameter and resembles a giant airship large enough to cover a great city like London or New York. What we would then see would be an object of unparalleled complexity and adaptive design. On the surface of the cell we would see millions of openings, like port holes of a vast space ship opening and closing to allow a continual stream of materials to flow in and out. If we were to enter one of these openings we would find ourselves in a world of supreme technology and bewildering complexity. Is it really credible that random processes could have constructed a reality, the smallest element of which – a functional protein or gene—is complex beyond our own creative capacities, a reality which is the very antithesis of chance, which excels in every sense anything produced by the intelligence of man? Alongside the level of ingenuity and complexity exhibited by the molecular machinery of life, even our most advanced artefacts appear clumsy.[13]

In comparison to a single cell, the Challenger spacecrafts are little toys for tots and NASA just a play station for boys and girls. A piece of humble pie, anyone?

DNA DYNAMICS

Experimental verifiable science has learned that in a human cell there are forty-six chromosomes containing about three billion base pairs of genes. This is the equivalent of over one thousand books of five hundred pages of genetically coded information. Anyone who can believe that this incredibly complex code, existing on a microscopic speck, is something that evolved on its own by chance, through time, is a fundamentalist religious materialist with a desperate faith in the non-existence of GOD.

Now if we multiply the amount of information in a single cell by the trillions of cells in one single human body, we get a picture of a miracle and a glimpse of the awesome intelligence that put it all together. There is more coded information in the cells of each of our bodies than in all the libraries and computers in the world.

Dr. Werner Gitt, the retired director and professor at the German Federal Institute of Physics and Technology (Physikalisch-Technische Bundesanstalt, Braunschweig), and the Head of the Department of Information Technology, wrote this about DNA:

> The cells of the human body can produce at least 100,000 different types of proteins, all with a unique function. The information to make each of these complicated molecular machines is stored on the well known molecule, DNA. We think that we have done well with human technology, packing information very densely on to computer hard drives, chips and CD-ROM disks. However, these all store information on the surface, whereas DNA stores it in three dimensions. It is by far the densest information storage mechanism known in the universe. Let us look at the amount of information which could be contained in a pinhead volume of DNA. If all this information was written into paperback books, it would make a pile of such books 500 times higher than from here to the moon! The design of such an incredible system of information storage indicates a vastly intelligent Designer.[14]

We can't out maximize or out miniaturize GOD'S creativity.

ONE PERSON'S JUNK IS ANOTHER'S TREASURE

Fifty years ago it was commonly believed that the DNA in our cells was simply a code for the creation of proteins, the building blocks of cellular life. It was then discovered that only about two to three percent of the DNA code was used to create proteins. At that point the prevailing scientific view and attitude was that the remaining ninety-seven percent was merely leftover junk from our evolutionary past. Thus, ninety-seven percent of our DNA was usually referred to as "junk DNA". Many believed this "fact" helped prove evolution to be true, and mocked the belief in creation. After all, if ninety-seven percent of our DNA is rubbish that has accumulated over hundreds of millions of years of evolution, then there simply is no room for creation. Creation scientists at this point basically responded with caution: don't jump to conclusions, wait until you understand more about the DNA code and you may be surprised to discover that most of it isn't junk. Their words have come to fulfillment. Most of the "junk DNA" that geneticists have experimentally examined over the past decade isn't trash at all. After studying portions of this ninety-seven percent of our DNA, they have come to realize that DNA is not just a string of a complex biological code creating proteins, but an amazingly complex operating system. Microsoft's Windows 7 is tiddlywinks in comparison.

Within this portion of the DNA, previously called junk, geneticists have discovered that there are a number of other codes. The codes presently identified include a cell memory code, a differentiation code, a replication code, and a splicing code, all integrated with each other.[15] There are hints of even more codes to be discovered as they examine these DNA portions in greater detail. Many of these codes are like instruction manuals that control how the many complex systems of the cell are created, function, interconnect and partner with one another. In an article called "Astonishing DNA Complexity Demolishes Neo-Darwinism," Alex Williams provides an insight into how miraculous our DNA actually is.

> The traditional understanding of DNA has recently been transformed beyond recognition. DNA does not, as we thought, carry a linear, one-dimensional, one-way, sequential code—like the lines of letters and words on this page. And the 97% in humans that does not carry protein-coding genes is not, as many people thought, fossilized 'junk' left over from our evolutionary ancestors. DNA information is overlapping-multi-layered and multi-dimensional; it reads both backwards and forwards; and the 'junk' is far more functional than the protein code, so there is no fossilized history of evolution. No human engineer has ever even imagined, let alone designed an information storage device anything like it. Moreover, the vast majority of its content is metainformation—information about how to use information. Meta-information cannot arise by chance because it only makes sense in context of the information it relates to...But recent discoveries show that so much information is packed into, on and around the DNA molecule that it is the most complex and sophisticated information storage system ever seen by mankind. No one ever imagined such a thing before, and we are still trying to understand the nature and depth of its information content.[16]

"Born Again", a blogger (contributor) on the Uncommon Descent website, a prominent Intelligent Design website that engages material evolutionists in a continual dialogue on the overwhelming evidence for design in our cosmos, references many articles and videos that reveal the incredible nano-technology that permeates every aspect of cellular life. Here are just a few of the resources he references:

Bioinformatics: *The Information in Life* – Donald Johnson – video
On a slide in the preceding video, entitled 'Information Systems In Life', Dr. Johnson points out that:

* the genetic system is a pre-existing operating system;
* the specific genetic program (genome) is an application;
* the native language has codon-based encryption system;
* the codes are read by enzyme computers with their own operating system;
* each enzyme's output is to another operating system in a ribosome;
* codes are decrypted and output to tRNA computers;
* each codon-specified amino acid is transported to a protein construction site; and
* in each cell, there are multiple operating systems, multiple programming languages, encoding/decoding hardware and software, specialized communications systems, error detection/correction systems, specialized input/output for organelle control and feedback, and a variety of specialized "devices" to accomplish the tasks of life.

Cells Are Like Robust Computational Systems, – June 2009
Excerpt: Gene regulatory networks in cell nuclei are similar to cloud computing networks, such as Google or Yahoo!, researchers report today in the

online journal Molecular Systems Biology. The similarity is that each system keeps working despite the failure of individual components, whether they are master genes or computer processors.... "We now have reason to think of cells as robust computational devices, employing redundancy in the same way that enables large computing systems, such as Amazon, to keep operating despite the fact that servers routinely fail."[17]

DO YOU THINK?

The human brain has been described as the most remarkable three pounds of matter in the entire universe. The human brain is estimated to have at least one hundred billion cells functioning together as a unit through trillions of interconnections.

Here are a couple of other amazing facts about the brain, which, by the way, is about seventy-five percent water:

> The number of internal thought pathways that your brain is capable of producing is: one followed by 10.5 million kilometers of standard typewritten zero's! Your brain is capable of having more ideas than the number of atoms in the known universe![18]
>
> Your brain contains about 100 billion neurons which is about sixteen times the number of people on Earth. Each of them links to as many as 10,000 other neurons. This huge number of connections opens the way to massive parallel processing within the brain. It is estimated that the human brain has a raw computational power between 10^{13} and 10^{16} operations per second. It is far more that 1 million times the number of people on Earth.[19]

How is it possible to believe that this incomparable three pounds of one hundred billion or more cells, each of which is a universe of complexity in itself, with trillions of electrical interconnections to each other, that enables you and I to live, love, experience, see, hear, feel, touch, contemplate our incredible universe and world, and relate and communicate with each other, is the result of fluke mutations and trial and error? This is astounding! It is somewhat understandable that people living 150 years ago, when Charles Darwin proposed the evolutionary theory, could have been deceived by it, since they had no comprehension of the incredible complexity of life. However, it is absolutely mind-boggling that people living in the twenty-first century, who are aware of this dynamic of life, could be beguiled by Darwin's mythology.

IRREDUCIBLY COMPLEX FEATURES IN ALL CREATURES

In his book *Darwin's Black Box,* biochemist Dr. Michael Behe, an intelligent design advocate, demonstrates the impossibility of Darwinian evolution by the concept of "irreducible complexity". An irreducibly complex structure or system is one which, in order to work, requires multiple parts to be present and functioning in harmony.

One of the examples Dr. Behe uses of irreducible complexity is the flagellum of a bacteria, which operates in similar manner to an outboard motor. For the flagellum to work, the various components of its motor-rotor, stator, bearing, hook, and filament-all need to be present and functioning in harmony. If any part was missing, or was not perfectly integrated with the other parts, the flagellum could not function. Most people can understand that every motor needs all of its basic parts to operate, and all of these components have to fit together or you have a motor that is a piece of junk.

One of the main tests of all legitimate scientific theories is whether or not they can be falsified by scientific data. One of the ways in which evolution's father, Charles Darwin, admitted that evolution could be falsified was through the discovery of irreducibly complex structures that could not reasonably have been produced through natural selection. Darwin wrote in *Origin of Species:* "If it could be demonstrated that any complex organ existed which could not possibly have been formed by numerous, successive, slight modifications, my theory would absolutely break down."[20]

Darwin wrestled with the issue of the eye being one such complex structure. The eye, of course, is incredibly more complex than Darwin could ever have imagined, not only in its physical organization and integrated coordination with the brain, but also in the genetic codes that provide the irreducibly complex information for the whole system of sight. Dr. H.S. Hamilton, in an article entitled: "The Retina of the Eye - An Evolutionary Road Block" comments:

> "That a mindless, purposeless, chance process such as natural selection, acting on the sequels of recombinant DNA or random mutation, most of which are injurious or fatal, could fabricate such complexity and organization as the vertebrate eye, where each component part must carry out its own distinctive task in a harmoniously functioning optical unit, is inconceivable... The total picture speaks of intelligent creative design of an infinitely high order."[21]

The obvious reality, in light of these new revelations of the cell, is that all of life is incredibly and irreducibly complex beyond Darwin's wildest imagination. The title of Dr. Behe's book, *Darwin's Black Box,* is a reference to the truth that to Darwin, and others living 150 years ago, the cell was a simple blob of protoplasm, like an empty black box. They had no comprehension of the miracle that a cell is.

Each cell is an engineering miracle, comparable to a city complete with billions of working residents, technologically advanced manufacturing plants, food and health distributors, sanitation systems, and a reproductive system. No matter how many billions of years a person can imagine, it would not be enough time for a single cell to come into existence by chance.

We realize today that many aspects of life are irreducibly complex. Each creature is a complete complex system which cannot function effectively without the integrated harmony between its multitudinous parts. From the cellular level on up, most parts in all living creatures cannot function on their own, but only in harmony and interdependence with many other parts of each organism, and with a complete complex integrated genetic system to create, control, repair and sustain the organism.

There are innumerable examples of irreducible complexity requiring intelligent design in living organisms. In his video series *Incredible Creatures that Defy Evolution,* Dr. Jobe Martin gives illustrations of different features of various creatures that are irreducibly complex, the result of an irreducibly complex intelligence. Dr. Martin was indoctrinated with evolution in a secular educational system, but in his book, *The Evolution of a Creationist,* he describes his transformation as he began to study a variety of animals that challenged the materialistic assumptions of his education.[22] His videos and books describe the fascinating world of animals to reveal scientifically sophisticated and irreducibly complex designs that cut the ground from underneath the evolutionary theory.

In their book, T*he Creation Explanation: a scientific alternative to evolution,* Dr. Robert Kofahl and Kelly Segraves reference the incredible integrated system of the worker honeybee:

> Every living creature is an integrated system which lives and functions as a whole. An excellent example of this is provided by the worker honeybee. Consider certain parts of the worker bee's body and their vital functions:
>
> 1. Compound eyes can analyze polarized light for navigation by sun in cloudy weather and for flower recognition.
> 2. Three single eyes on the head probably have some navigation function.
> 3. Antennae supply sense of smell and touch.
> 4. Grooves on the front legs clean antennae.
> 5. Hairs on head, thorax, and legs collect pollen.
> 6. Pollen baskets on rear legs carry collected pollen.
> 7. The tube-like proboscis sucks nectar, honey, water, curls back under head.
> 8. Mandibles crush and form wax for comb-building.
> 9. A honey sac provides temporary storage of honey.
> 10. Enzymes in honey sac begin transformation of nectar to honey.
> 11. Glands in the abdomen produce beeswax, which is secreted as scales on rear segments of body.
> 12. Long spines on middle legs remove wax scales from glands.
> 13. Pronged claws on each foot cling to the flowers.
> 14. Glands in head of adult worker make royal jelly for the development of a new queen.
> 15. Marginal hooks fasten front and rear wings together for flight, disengage at rest for compact storage of wings.
> 16. Barbed poison sting serves for defense.

17. Complex instincts cause entire hive to function as a single organism.

Without all of these design features and many more the worker bee could not function in the hive; as a result the entire hive would perish.[23]

In his book *Billions of Missing Links,* Dr. Geofrey Simmons reveals the irreducible complexity of numerous creatures that have no evolutionary explanation, apart from blind faith that they all just happened through fluke mutations, natural selection, and the magical kiss of time. He writes:

"How complex is too complex for the theory of evolution to explain? The adult blue whale has 100 quadrillion cells, and each cell has up to one billion chemical compounds. How fast is too fast to be an accident? Enzymes work within cells at a millionth of a second. The impact of light on the retina to create vision takes 200 femtoseconds. A femtosecond is a millionth of a billionth of a second. When does a complicated life cycle speak against chance occurrence or a lucky series of mutations? There are known parasites that require at least two unrelated and distant hosts to complete one life cycle. There are insects that depend on protozoa in their stomach to survive, which in turn depend on even smaller microorganisms attached to them for survival. How impossible should a body part be? The lens of the human eye has 1000 layers of transparent, living cells."[24]

The reality is that every kind of creature is irreducibly complex, based not only on its uniquely designed features but on the integration of all of its DNA-based codes. We can exist without our hands, feet, eyes, ears, and a few other parts of our bodies, but we cannot exist without the integration of all of our DNA codes at the cellular level. The same is true for every other creature. The only reasonable explanation is that every creature and the codes for every creature are the result of super intelligent design. The rest is history: the reproduction of a variety of species adapted from the original DNA code to the changing environments.

I believe that doctrinaire evolutionists refuse to accept the concept of "irreducible complexity" because they fear that every form of life has aspects that are "irreducibly complex" and cannot be explained by natural selection or mutations. All of the incredible intricacies of life not only falsify the theory of evolution by way of irreducible complexity, but clearly testify to an awesome intelligent Designer.

Summing up his conclusions on the state of the evolutionary theory in the light of the complexity of life, molecular biologist Dr. Michael Denton wrote "Ultimately, the Darwinian theory of Evolution is no more nor less than the great cosmogenic myth of the twentieth century."[25] The well known philosopher, satirist, and author Malcolm Muggeridge came to a similar conclusion, writing "The theory of Evolution... will be one of the great jokes in the history books of the future. Posterity will marvel that so flimsy and dubious a hypothesis could be accepted with the incredible credulity it has."[26]

SIGNS OF INTELLIGENCE

If something is created by intelligent design, it bears the marks of intelligent design. Cars, computers, planes, and everything humans have created carry the marks of intelligent design, although not always as intelligent as we would like. Likewise, it is written: ***"For what can be known about GOD is plain to them, because GOD has made it plain. For*** *ever since the creation of the world GOD'S invisible qualities, His eternal power and divine nature have been clearly seen, being understood from what has been made, so that men are without excuse." (Romans 1:19,20)*

You can't put the designer into a test tube, but you can infer from the marks of design that intelligence was involved. In his book, *Signature of the Cell,* Dr. Stephen C. Meyer demonstrates how the digital codes of the cell are a clear signature of intelligent design. In fact, the codes reveal not just human intelligence but a super intelligence beyond our human comprehension.

It is astounding that materialists reject a miracle-working GOD, but believe that, through the magic of time, frogs or other slimy inhabitants came from nothing and were transformed into princes. If the precise tuning of the cosmos, the DNA codes and operating system, the incredible nano-technology of the cell, the computational ability and electrical connectivity of the brain, and the irreducible complexity of all of life are not enough to convince people of the impossibility of evolutionary chance processes producing these organizational marvels, nothing ever will. They are without excuse.

How much evidence does a person need? The entire cosmos and every plant and animal is a miracle: something that humanity cannot do. **JESUS said: "What is impossible for man is possible with GOD!"(Matthew 19:26).**

Despite all of our claimed scientific knowledge and technological achievements, humanity cannot create a single cell or a blade of grass. We can deform and destroy bacteria, but we can't turn it into any other form of life. Once a bacterium, always a bacterium! Nevertheless, evolutionists stridently proclaim that everything happened by chance and trial and error through the magical kiss of time. Go figure!

Recently the world-renowned cosmologist Stephen Hawking, in his new book, The Grand Design, proclaimed that GOD had nothing to do with the creation of the universe, but rather that the laws of physics created the universe all by themselves. However, he never got around to explaining exactly where the laws the physics came from. Do the laws of physics exist where there is no universe?

Professor Richard Lewontin, a geneticist and one of the world's leaders in evolutionary biology, wrote this very revealing statement that highlights the implicit philosophical bias against GOD that he, Stephen Hawking, and numerous other scientists hold:

> We take the side of science in spite of the patent absurdity of some of its constructs, in spite of its failure to fulfill many of its extravagant promises of health and life, in spite of the tolerance of the scientific community for unsubstantiated just-so stories, because we have a prior commitment, a commitment to materialism. It is not that the methods and institutions of science somehow compel us to accept a material explanation of the phenomenal world, but, on the contrary, that we are forced by our a priori adherence to material causes to create an apparatus of investigation and a set of concepts that produce material explanations, no matter how counter-intuitive, no matter how mystifying to the uninitiated. Moreover, that materialism is an absolute, for we cannot allow a Divine Foot in the door.[27]

Not only is the Divine Foot in the door, but it continues to trample on their sacred secular mythologies. Fortunately, there are many scientists and others who, when faced with the googolplex of complexity of the cosmos and of all life, are, unlike Drs. Hawking and Lewontin, turning their thoughts to their Creator. The excellent web-site, Creation-Evolution Headlines, one of the very best on the internet for up-to-date information on this issue from a Christian Biblical perspective, has this posting from October 29, 2007 regarding Sir Anthony Flew, one of the most well known atheists of the twentieth century:

> Book: Intelligent Design Argument Turns Leading Atheist to God 10/29/2007
> "There is a God," announces a former leading atheist on the cover of his new book. Antony Flew changed his mind a few years ago partly because of the design argument... In an exclusive interview with Benjamin Wiker on To the Source, Antony Flew made it clear that intelligent design was a decisive factor in his change of heart:
> Sir Anthony Flew: "There were two factors in particular that were decisive. One was my growing empathy with the insight of Einstein and other noted scientists that there had to be an Intelligence behind the integrated complexity of the physical Universe. The second was my own insight that the integrated complexity of life itself – which is far more complex than the physical Universe – can only be explained in terms of an Intelligent Source. I believe that the origin of life and reproduction simply cannot be explained from a biological standpoint despite numerous efforts to do so. With every passing year, the more that was discovered about the richness and inherent intelligence of life, the less it seemed likely that a chemical soup could magically generate the genetic code. The difference between life and non-life, it became apparent to me, was ontological and not chemical. The best confirmation of this radical gulf is Richard Dawkins' comical effort to argue in The God Delusion that the origin of life can be attributed to a "lucky chance." If that's the best

argument you have, then the game is over. No, I did not hear a Voice. It was the evidence itself that led me to this conclusion."[28]

THE GOD OF MIRACLES!

Nothing is natural in the sense that it exists in and of itself. We call the material realities of life natural only because they exist in nature and not because they exist apart from GOD'S intelligent design, power, or will. The truth is that every atom, every molecule, every cell, every life form exists only through the incredible wisdom and power of GOD, not only in the process of creation, but also in its maintenance. One of the lamentable side effects of scientific materialism is that, although people are surrounded by miracles every day, their materialistic blinders prevent them from seeing. Life is breathtaking for those who recognize the miracles of GOD that encircle us, including the miracle of our own lives! Life is one incredible miracle after another! "I was blind, but now I see."

It is time to sign up and join the Heavenly choir, the cosmos chorus, and GOD'S people in praising and glorifying our awesome Creator. Make it so!

Psalm 96

Sing to the LORD a new song;
sing to the LORD, all the earth.
Sing to the LORD, praise his name;
proclaim His salvation day after day.
Declare His glory among the nations,
His marvelous deeds among all peoples.
For great is the LORD and most worthy of praise;
He is to be feared above all gods.
For all the gods of the nations are idols,

but the LORD made the heavens.
Splendor and majesty are before Him;
strength and glory are in His sanctuary.
Ascribe to the LORD, O families of nations,
ascribe to the LORD glory and strength.
Ascribe to the LORD the glory due His name;
bring an offering and come into His courts.
Worship the LORD in the splendor of His holiness;
tremble before Him, all the earth.
Say among the nations, "The LORD reigns."
The world is firmly established, it cannot be moved;
He will judge the peoples with equity.
Let the heavens rejoice, let the earth be glad;
let the sea resound, and all that is in it;
let the fields be jubilant, and everything in them.
Then all the trees of the forest will sing for joy;
they will sing before the LORD, for He comes,
He comes to judge the earth.
He will judge the world in righteousness
and the peoples in His truth. AMEN and AMEN!

Chapter 2

GOD vs. Nothing: Why Evolution Flunks Logic, Math, and Science

"In the beginning was the Word, and the Word was with GOD, and the Word was GOD. He was with GOD in the beginning. Through Him all things were made; without Him nothing was made that has been made. In Him was life, and that life was the light of men. The light shines in the darkness, and the darkness has not overcome it." (John 1:1-5)

ORIGINS: GOD OR NOTHING?

Origins: where did it all come from? In *On The Origin of Species by Means of Natural Selection, or The Preservation of Favored Races in the Struggle for Life,* published in 1859, Charles Darwin presented a new idea of where species, and ultimately life, originated from. Quite understandably, in the historical and contemporary writings of humanity, it has always been a matter of great interest as to where we and everything around us originated. But why? Other than curiosity what difference does it make?

Dr. Henry Morris, one of the fathers of the modern day creation science movement, in his scientific and devotional commentary on Genesis entitled *The Genesis Record* wrote:

> "The future is bound up in the past. One's belief concerning his origin will inevitably determine his purpose and his destiny. A naturalistic, animalistic concept of beginnings specifies a naturalistic, animalistic program for the future. An origin at the hands of an omnipotent, holy, loving GOD, on the other hand, necessarily predicts a divine purpose in history and an assurance of the consummation of that purpose."[29]

The origin of anything and everything, from the cosmos to the milk carton, is directly connected to its purpose, value or reason for existence. If something has its origin in random chance or fluke then it has no intrinsic purpose or value. The exact opposite is the case for everything that is intentionally brought into existence with a plan or purpose: its essential value is in the fulfillment of that purpose. Although it may be co-opted for another purpose, like a newspaper being used for lighting fires, it wasn't created for that reason. For all matter without a rational mind and soul, the issue is non-existent. However, this is not the case for us humans. We do have a mind which GOD has endowed us with to contemplate life, including our origin.

Darwin's answer to the question of origins was wrong, but it was a critical question in his life that he recognized was central to understanding the quintessential nature of human life. It is tragic that he chose to believe that human life, including his own, was a mere accident of nature rather than the purposeful action of His Creator. The rest is history, and much tragedy in that history, as we will see in chapter 7.

The question of origins is of ultimate importance for each individual, whether he or she recognizes it or not. The common understanding of our origin will be directly reflected in a society's world view and consequent values. Avoidance of the issue doesn't mitigate its impact on our life or world. Many may stick their heads in the sand like ostriches, and go about their life thinking that where they came from doesn't make any difference to how they live; however, beliefs have consequences, and no other belief has greater consequences than the belief about origins. Whatever someone believes about his or her own genesis and the genesis of the universe and world will have tremendous implications for his or her life and have a ripple effect on the life of everyone. No one is an island: we affect our world, and our world impacts us.

So where did we come from? There are two fundamentally opposing answers to the question of origins that are common today. The answer which has existed from the beginning is that everything, including humanity, is the result of the creative work of a GOD above and beyond us. This is the information communicated in Genesis and affirmed repeatedly throughout the Bible.

The first truth that GOD reveals in His message to us is our genesis because it is foundational to all other truth. In the opening words of the Bible, GOD testifies that He is the Creator of our universe and all the mysteries contained therein. In microscopic cells, in vast galaxies and, perhaps most clearly, in the miracle of our own lives and bodies, the awesome reality of our Creator GOD is revealed. Dr. Henry Morris lists many realities that Genesis reveals the origin of, including the universe, order and complexity (Genesis 1:2 formlessness before creation), the solar system, the atmosphere and hydrosphere, life, humanity, marriage, evil, language, government, culture, nations, and religion.[30] Every segment of life is impacted by our understanding of our origin.

The second answer to the question of origins, which has been growing in popularity over the past century and a half, is the belief in biological and cosmic evolution. This materialistic-based belief is that everything mysteriously came out of nothing. However, in its inflexible insistence on explaining everything, including our origins, on the basis of material elements alone, scientific materialism has sucked itself into a vacuum of nonsense. Materialistic mythology begins with nothing, and then through the power of a random explosion of that nothing—space, time, matter and the transcendent laws of the universe magically, without a magician, appear out of nowhere. Materialists believe that this is followed by cosmic and biological evolution, where material atoms luckily organize themselves into all the wonders of the universe and of life itself.

In cosmic and biological evolution, everything just happens out of nothing without any purpose or planning. The universe just happened. Life just happened. The sparrow and the rainbow just happened. The one-hundred-quadrillion-celled blue whale, with a billion or more precisely arranged molecules in each of those cells, just happened. The human mind with its trillions of electrical connections just happened. It all happened out of nothing through the magical kiss of chance and time.

Only by using materialistic evolutionary math can you multiply nothing times nothing and get everything we have in the present. In this secular materialistic theory for the origin of everything, in the end you are back to nothing. Many believe that eventually the universe itself will implode or disintegrate in some fashion. This rationalizes the materialistic *"eat, drink, and live by your selfish genes for tomorrow you die"* approach to life by those who prefer the evolutionary mythology. In the end their math all works out when everything becomes nothing, but meanwhile back in reality everything is here. What is amazing is that so many of the evolutionists scoff at Christians and others for believing in miracles. At least we know the Almighty Miracle Worker; all they have is nothing.

Our generation often looks contemptuously at the ancient peoples who created their own idols to worship and to give some explanation to the confusing circumstances of life. Although these peoples made idols in their own often depraved image, at least they comprehended that everything had come from something beyond them, that there were realities greater than themselves in the universe. Tragically, having rejected the living GOD, they were left to live in slavery in the haunts of their imaginary idols. While people today scorn these ancient peoples for their foolish idols, many in our generation have fallen for an even greater absurdity, namely, that everything came from nothing all by itself. In the delusion of materialistic thinking fostered by evolutionism, many have developed their own math where nothing times nothing equals everything. Go figure!

It is written: *"I look on the faithless with pity for they do not obey Your Word" (Psalm 119:158).* Despite all that GOD has given to people, many face lives of existential hollowness and despair. Without faith in GOD and His Word, they fearfully, or stoically, or under the influence of intoxicants, walk towards the cliff at the end of their road. They never know when the sandy ground underneath them will collapse, creating the chasm into which they fall. There is a chasm or valley of death for each of us when life in this world ends, which could be at any time. Those who trust in CHRIST know there is a bridge over the chasm and into His eternal realm. GOD invites people to come to the bridge which is CHRIST, the perfect GOD-Man. His cross is the bridge between life and eternity. *JESUS said: "I AM the Way and the Truth and the Life, no one can come to the FATHER except through Me"* **(John 14:6).** Life does not have to be a vacuum, from nothing to nothing; rather, through trusting in and following the SON, life can be a grand march to an overflowing eternal life in the presence of GOD.

PHILOSOPHICALLY LUCKY MATH OR AN INCREDIBLE CREATOR?

Math wasn't Darwin's forte, and it isn't the forte of evolutionists today. In the first chapter we caught a glimpse of the fantastic fine-tuning involved in the dynamics of the universe. The mathematical probability of the universe being so precise in so many different parameters is astronomically unlikely. In a similar vein, the probability of life developing through the laws of chemistry is infinitesimal, non-existent for all sober people. In addition, there is the vast number of positive interrelated mutations that would be necessary for one creature to be transformed into another kind of creature, and this would have to be repeated for all the life forms all the way through the evolutionary process. There is no limit to their imaginations, but they flunk real math.

The only response evolutionists have to these mathematical impossibilities of life is that we just got lucky. Their math is philosophically lucky math; lucky like someone winning the lottery repeatedly day after day for billions of years. Most lotteries are a one in a million chance, whereas the chance of the cosmos and life evolving on their own is less than one in a googolplex. No, we aren't lucky - we are blessed with a GOD Who Is Who He Is, an absolutely amazing Creator and FATHER.

INFORMATION ISN'T MATTER, IT IS MIND OVER MATTER

At the beginning of the twenty-first century we are living in what is now commonly called "the information age". Through telecommunications and other computerized systems, we are surrounded by all kinds of information: some good, some bad, some trivial, some pivotal. But what is information? It isn't material. Information may be data about material or about those things that are immaterial, but it isn't material. It can be communicated with the use of material, but it isn't material. Apart from a mind, information is non-existent. Information, like beauty, is only in the mind: a mind that can perceive and interpret it.

Dr. Phillip Johnson, a retired professor of law at the University of California Berkeley, the author of *Darwin on Trial,* and one of the founders of the Intelligent Design Movement, in his book *Defeating Darwinism by Opening Minds* explains the difference between material and information. He wrote:

> This book you are reading, like any other, contains information written on paper with ink. The information did not always take that physical form, however. Originally I wrote it on a word processor, and it existed only as an electronic file on a computer disk. I sent some completed chapters by email to friends and colleagues for criticism. The information in each chapter was exactly the same whether it was recorded on paper or on a computer disk in some fragmented and disembodied form as it moved over the links of the Internet.
>
> Information is also stored by some poorly understood means in our brains. If all copies of Shakespeare's plays were destroyed, nothing would

be permanently lost. Actors who had learned the roles could easily re-create the texts from memory.

Such examples tell us that information is an entirely different kind of stuff from the physical medium in which it may temporarily be recorded. It would be absurd to try to explain the literary quality or meaning of a book as an emergent property of the physical qualities of its ink and paper. The message comes from an author; ink and paper are merely the media. Similarly, the information written in DNA is not the product of DNA. Where did the information come from? Who or what is the author?[31]

Dr. Werner Gitt has written many scientific papers in the field of information science. In his book, *In the Beginning was Information,* he wrote: "It should now be clear that information, being a fundamental entity, cannot be a property of matter, and its origin cannot be explained in terms of material processes."[32]

Most computers are loaded with all kinds of information but it is not information to the computer itself. It is only information to those with a mind to operate a computer and to input and interpret its data. The words on this page are not information to a dog or a monkey. The ringing of a bell may be information to a trained hungry dog, but a written invitation to a banquet is not going to turn on the saliva. Information always requires a mind to create it and a mind to interpret it.

Information is contra the random, mindless process of evolution. The DNA code is a preprogrammed incredibly complex library of information which requires an incredibly complex mind to create it and an incredibly complex mind to understand it. As we shared in chapter one, this preprogrammed complex information in all life forms is the signature of GOD.

In mathematics, the almost pure science, it is a statistical impossibility for the genetic code to have evolved. Dr. A.E. Wilder-Smith had three doctorates, his first being in physical organic chemistry at Reading University, England in 1941. After becoming a full professor at the University of Geneva, he earned a second doctorate in pharmacology there and later, a third in pharmacological sciences at ETH, a senior university in Zurich, Switzerland. He was a committed Christian who believed and upheld the Genesis account of history in many debates and presentations. As I recall, he occasionally used the example that the probability of DNA occurring on its own is less than that of someone randomly throwing out 100,000 white cards with letters on, from an airplane at 6000 feet, and having them land beside each other in such a way as to spell out his name.

Sir Fred Hoyle, the famous British astronomer and one time atheist, was shocked into reality when he realized the incredible complexity of the universe and of life, and of the complete impossibility of it happening by chance. He enraged many of his scientific colleagues by pointing out repeatedly how irrational it was to believe that random chance could produce the remarkable reality of life.

> Hoyle compared the random emergence of even the simplest cell to the likelihood that 'a tornado sweeping through a junk-yard might assemble a Boeing 747 from the materials therein.' Hoyle also compared the chance of obtaining even a single functioning protein by chance combination of amino acids to a solar system full of blind men solving Rubik's Cube simultaneously. Published in his 1982/1984 books Evolution from Space (co-authored with Chandra Wickramasinghe), Hoyle calculated that the chance of obtaining the required set of enzymes for even the simplest living cell was one

in $10^{40,000}$. Since the number of atoms in the known universe is infinitesimally tiny by comparison (10^{80}), he argued that even a whole universe full of primordial soup would grant little chance to evolutionary processes.[33]

It doesn't matter how many billions of years you allow tornadoes to sweep through junk yards, or blind men to play with Rubik's cubes, or people in airplanes to throw out letters, you will never end up with complex creative order of any kind. Boeing 747's require not only time but an intelligent designer.

Dr. Gary Parker is another atheistic evolutionist turned creationist. In his book *Creation: Facts of Life: How Real Science Reveals the Hand of GOD* he shares how he used to teach evolution and enjoyed ridiculing Christian creationists. He wrote:

> For me evolution was much more than just a scientific theory. It was a total world-and-life view, an alternative religion, a substitute for God. It gave me a feeling of my place in the universe, and a sense of my relationship to others, to society, and to the world of nature that had ultimately given me life. I knew where I had came from and where I was going.
>
> I had heard Christians and other "religious fanatics" talk about "back to God, back to the Bible, back to this, or back to that." But for me as an evolutionist, the best was yet to come. And, as a scientist and professor of biology, I could help make it happen. By contributing to the advances in science and technology, both directly and through my students, I could be part of the process of bringing "heaven on earth".

Let's face it. Evolution is an exciting and appealing idea! A lot of scientific evidence can be used to support it. Perhaps most importantly for me and many others, evolution means there is no God, no "Creator" who sets the rules. Human beings are the top. Each of us is his or her own boss. We set our own rules, our own goals. We decide what's best for us.

I didn't just believe evolution; I embraced it enthusiastically! And I taught it enthusiastically. I considered it one of my major missions as a science professor to help my students rid themselves completely of old, "pre-scientific" superstitions, such as Christianity. In fact, I was almost fired once for teaching evolution so vigorously that I had a Christian students crying in my class![34]

Dr. Parker's mind and heart were changed as he came to know CHRIST and began to clearly see how all of life was intelligently designed. Dr. Parker's book *Creation: Facts of Life: How Real Science Reveals the Hand of GOD* is one that I highly recommend for it provides a great overview of the issues and the evidence for creation in a very readable format for most non-scientists. It has been a great help to me to understand many of the basic truths involved in the creation-evolution conflict.

On the necessity of information for all life forms, Dr. Parker wrote:

Did you ever wonder what makes an airplane fly? Try a few thought experiments. Take the wings off and study them; they don't fly. Take the engines off, study them; they don't fly. Take the pilot out of the cockpit; the pilot doesn't fly. Don't dwell on this the next time you're on an airplane, but an airplane is a collection of non-flying parts! Not a single part of it flies! What does it take to make an airplane fly?... Creative design and organization.[35]

As GOD reveals in the Bible and as Dr. Parker points out in his book, you and I are made out of dust, the same chemical and mineral molecules that we walk on every day and that are used to create airplanes and everything else. Not a single molecule in our bodies is alive, but we are alive. What is the difference? Our bodies have creative design and organization, and the breath of GOD. The breath of GOD is the difference between us and an airplane. GOD'S breath makes us living, thinking, spiritual beings rather than just molecular machines.

Every one of the billions of molecules in each of our cells needs to be in a very specific place and guided by a software program to accomplish a very specific task so that each cell can function, and so that we can live. Through creative design and organization GOD has taken the dust of the earth and created bodies for you and me. He did it once, and He has promised through CHRIST that He will do it again when He creates new eternal bodies for those who trust in Him. Praise GOD for these temporary bodies and the eternal ones to come!

The information required for the production and assembly of a Boeing 747 is miniscule in comparison to the information required for the creation of a single cell. Evolution would have to involve not just one magical tornado but millions of them on every branch up the evolutionary tree. To believe in evolution, you have to believe in trillions of miracles that just happen by themselves; magic without a magician. Inevitably, evolutionists say that you just need more time. Time is the magic wand they wave and "kazaam", life is transformed from a toad into a prince.

THE FACTS OF LIFE VS THE MYTHS OF LIFE

Where does life come from? One verifiable scientific fact of life is that life comes only from life. It is a statement of materialistic faith when evolutionists state that life originated out of pond scum or the vents of ancient volcanoes through chance occurrence. There isn't a shred of scientific evidence to suggest that this happened; it is pure conjecture based on faith in material atoms.

Once upon a time it was an article of faith held by many scientists that life could arise spontaneously out of the elements. Dr. Carl Werner, in his meticulously researched and illuminating book, *Evolution: The Grand Experiment,* has an excellent chapter on the history of the scientific belief in spontaneous generation. I encourage everyone who desires a good clear understanding of the limitations of science and the shoddy foundation of the evolutionary theory to read Dr. Werner's excellent book. On his website promoting the book we are given an introduction to its background:

> In his sophomore year of college, Dr. Carl Werner was challenged by a fellow classmate with these words: "I bet you can't prove evolution." Disturbed by his classmate's somewhat poignant criticisms of the theory, Dr. Werner (a biology major at the time) pondered this conversation for some time. From this point, he began his quest for an answer. Driven by his stringent scientific background and exceedingly curious nature, Dr. Werner read and researched every topic he could find on evolution, including geology, biology, palaeontology, biochemistry and cosmology. After 18 years of study, Dr. Werner considered himself ready to begin a series of experiments by which to test evolution. For the next 10 years, 1997-2007, Dr. Werner traveled to the best

museums and dig sites around the globe photographing thousands of original fossils and the actual fossil layers where they were found.[36]

In his chapter on life's origin, Dr. Werner reviews the scientific achievements of Dr. Louis Pasteur who proved through scientific experimentation that life does not spontaneously generate. Although it is a fact of science that life does not spontaneously come into existence, materialistic scientists continue to believe and teach that life just happened as a result of chance chemical processes. In many, probably most, high school and university biology textbooks, abiogenesis, the belief that life came into existence by spontaneous generation through material alone, is upheld and promoted even though there is no experimentally verifiable science that establishes it.

Evolutionary biologists often say that their theory doesn't have anything to do with the actual origin of life but only the development of life from the first cell onwards. However, if the evolutionary theory is not capable of getting to the starting gate of life with the creation of a first cell, it collapses before it even begins.

The practical reality, of course, is that it is no easy process to create a cell. Despite all of the scientific experiments and contemporary understanding of the cell, no scientist has been able to create a cell or even come within a million miles of doing so. Nevertheless, evolutionism believes that a simple cell of some kind got lucky and formed itself through chance processes. I recall the excitement in my university introductory biology class in the mid-seventies over a relatively new book called *Biochemical Predestination* by Dr. Dean Kenyon, a convinced evolutionist. At that time Dr. Kenyon claimed that it was inevitable for life to arise out of the chemical soup of early earth. However, the reality of the complexity of life eventually got to him.

Kenyon received a BSc in physics from the University of Chicago in 1961 and a Ph.D. in biophysics from Stanford University in 1965. In 1965-1966 he was a National Science Foundation Postdoctoral Fellow in Chemical Biodynamics at the University of California, Berkeley, a Research Associate at Ames Research Center. In 1966, he became Assistant Professor of Biology at San Francisco State University. He has been Emeritus at San Francisco State University since 2000. In 1969, Kenyon and coauthor Gary Steinman published Biochemical Predestination, a book on the origins of life advocating a theory of natural chemical evolution. Kenyon's views changed around 1976 after exposure to the work of young-earth creationists. In his own words,

"Then in 1976, a student gave me a book by A.E. Wilder-Smith, The Creation of Life: A Cybernetic Approach to Evolution. Many pages of that book deal with arguments against Biochemical Predestination, and I found myself hard-pressed to come up with a counter-rebuttal."[37]

Dr. Kenyon opened his eyes and realized that, with the increasing complexity of the cell, belief in biochemical predestination was absurd. Even the most basic cell involves millions of molecules, all in an organized arrangement working at various tasks in harmony with other molecules fulfilling diverse functions. A so-called "simple cell" has a much greater organized complexity than the most sophisticated computer system in the world. There are none so blind as those who refuse to see. Thankfully, there are many like Dr. Kenyon who are opening their eyes and their minds to the awesome reality of GOD and aren't ashamed to tell others.

ARE YOU RELATED TO THE APES?

Evolutionists, especially in the secular media, are very fond of trotting out the false statistic that humans and apes are about ninety-eight or ninety-nine percent similar in their genetic composition. Although the statistic is still tossed around, it has little support in terms of the newest and best scientific studies today. It may have seemed to be true thirty years ago, but it isn't factually accurate today, as many articles by both creationists and non-creationists have pointed out. The Creation-Evolution Headlines website had this article on its June 29, 2007 page:

> The Chimp-Human 1% Difference: A Useful Lie
> Jon Cohen made a remarkable admission in Science this week. The popular notion that humans and chimpanzees are genetically 99% similar is a myth, and should be discarded. Since 1975, textbooks, the media and museums have emphasized this close similarity; but now, Cohen quoted a number of scientists who say the number cannot possibly be that small and probably cannot be quantified. Since the statistic has outlived its usefulness, it should be discarded.
> The original claim by Allan Wilson in 1975 came from studies of base substitutions when genes were compared side by side. Other comparisons, however, yield very different results. Human and chimp genomes differ markedly in:
>
> - Chunks of missing DNA
> - Extra genes
> - Number of chromosomes and chromosome structure
> - Altered connections in gene networks
> - Indels (insertions and deletions)

- Gene copy number
- Coexpressed genes

In this last measure, for instance, a 17.4% difference was found in genes expressed in the cerebral cortex. Cohen recalled the December 2006 paper from PLoS One where Matthew Hahn found a "whopping 6.4%" difference in gene copy numbers...But even that number is misleading. Different measures produce such different results, it is probably impossible to come up with a single percent difference that wouldn't misrepresent the picture. Scientists are not sure how to prioritize the measures to study, because "it remains a daunting task to link genotype to phenotype." Sorting out the differences that matter is "really difficult," said one geneticist. A stretch of DNA that appears meaningless may actually be vital for gene regulation...At the end of the article, Cohen quoted Svante Paabo, who said something even more revealing. After admitting he didn't think there was any way to calculate a single number, he said, "In the end, it's a political and social and cultural thing about how we see our differences." Jon Cohen, News Focus on Evolutionary Biology, "Relative Differences: The Myth of 1%," Science, 29 June 2007: Vol. 316. no. 5833, p. 1836, DOI: 10.1126/science. 316.5833.1836.[38]

A more recent "Study found only 86.7% genetic similarity when segments of human and chimpanzee DNA (totaling 1,870,955 base pairs) were laid side by side. This study also included indels (insertions/deletions) in addition to substitutions." ref: Tatsuya Anzai st al., "Comparative Sequencing of Human and Chimpanzee MHC Class | Regions Unveils Insertions/Deletions As the Major Path to Genomic

Divergence," Proceedings of the National Academy of Sciences, USA 100 (2003); 7708-13[39]

The original "guesstimates" that evolutionists used were based on a comparison of only a small portion of the genomes of chimps and humans. There is much new data that indicates a less than ninety-five percent similarity in our DNA sequencing with that of chimps. However, as the above articles indicate there is still much work that needs to be done in comparing the entire chimp genome to that of humans, and still much mystery in the functioning of the DNA, particularly in the ninety-seven percent portion which they used to call "junk DNA". Consequently, the figure needs to remain fluid, subject to many more studies. The boasting of the ninety-eight to ninety-nine percent figures in evolutionist educational and media materials is old science and nothing more than a wish fulfillment projection.

There are two important factors that evolutionists don't usually mention with regard to DNA similarities. Firstly, even if there were only a one percent difference in the genes between humans and chimps, that would be equal to about thirty books of five hundred pages of three hundred words per page of genetically coded information (3 billion letters X 1% = 30 million letters/6 letters per word= 5 million words/300 words per page=16,666 pages/500 pages per book=33 books). That is an amazing amount of complex genetic code that needs to be added, subtracted, or changed by random mutations. If the actual figure is closer to ninety percent similarity, then the difference would be more like three hundred books or forty-five million words. Secondly, it is important to note that our DNA is also apparently about eighty to eighty-five percent similar to that of mice,[40] and about seventy-five percent similar to a nematode,[41] which is little worm that lives in the dirt. No evolutionist is claiming that mice or nematodes are seventy-five percent human, or that we are seventy-five percent mouse or nematode! Humans also have a lot of similarities to pigs in our digestive tracts, nutritional needs, hearts and skin. Can you say "oink, oink"?

DNA is the genetic building material that GOD designed, engineered, and created, and through which He has formed all of the different creatures on earth. As many creation scientists have pointed out, just as human chemists, engineers, architects, and carpenters use the same basic building materials in building homes, skyscrapers, cars, airplanes and computers, so GOD uses the same dust of the earth and the same basic DNA code to create the incredible variety of life on earth. The difference is not in the material used but the intelligent creative design for which the materials are used. Even food materials like ice-cream are sometimes created from edible oil products, and we will not mention some of the other ingredients that food manufacturers put into processed foods these days. You can take oil and gas and use them for operating vehicles, flying planes, or for your supper tonight. It all depends on the intelligent creative use of the materials. The same is true with DNA.

Apes and humans do share some similarities, so it isn't surprising at all that GOD used some of the same basic DNA codes to form apes and humans. We have some similarities with pigs and so it shouldn't surprise us that GOD used similar DNA in those areas that we have in common. Are we related to apes and pigs? Not by descent, but we do have the same Creator who used the same dust of the earth, the same chemical molecules, and some of the same DNA codes to create us and everything else in our world. What an awesome intelligent creative Designer!

MUTATIONS: THE ENGINE OR THE END OF EVOLUTION?

One of the great obstacles for the evolutionists is how to explain the creation of new information to account for the development of new features and new creatures. According to the evolutionary theory, where does all of this incredibly complex information come from? Evolutionism holds that all life has evolved from a single cell through the power of natural selection and random mutations. The process of natural selection, by definition, doesn't create any new genetic material but merely selects between different traits that are already expressed. There is no controversy whatsoever with the scientific reality of natural selection. It is very observable in nature today. It is verifiable experimental science. No creation scientist that I have met or read rejects natural selection. Natural selection is not evolution in the broad sense, but the interaction of created forms of life with a changing and decaying environment. Natural selection does not transform one kind of creature into another. It provides for the variation we see within species, such as different breeds of dogs and cats.

The only mechanism that evolutionism offers for the creation of new genetic material to produce new traits or creatures is random genetic mistakes called mutations. Mutations are proclaimed to be the powerhouse or engine of evolution that drives the creation of new genetic information. Dr. Theodosius Dobzhansky, one of the foremost evolutionists of the twentieth century wrote, *"The process of mutation is the only known source of the raw materials of genetic variability, and hence of evolution"* (American Scientist, Winter 1957, p. 385).[42]

Charles Darwin lived prior to the understanding of genetics and mutations. He was a great story teller with a vivid imagination who developed all kinds of tales to explain how various animals were able to be transformed into different creatures. Believing that whales may have evolved from black bears, he wrote:

> In North America the black bear was seen by Hearne swimming for hours with widely opened mouth, thus catching, like a whale, insects in the water. Even in so extreme a case as this, if the supply of insects were constant, and if better adapted competitors did not already exist in the country, I can see no difficulty in a race of bears being rendered by natural selection, more and more aquatic in their structure and habits, with larger and larger mouths, till a creature was produced as monstrous as a whale.[43]

There wasn't any evidence to suggest that bears evolved into whales, but that didn't stop Darwin from imagining it to be so. This kind of story-telling speculation that Darwin used is still the *modus operandi* of evolutionists today, although now they explain these changes as being the result of genetic mutations and not behavior.

Today, one of the proposed possibilities that evolutionists imagine is that whales may have evolved out of some hyena type creature. What are the mathematical possibilities of this happening? Dr. Carl Werner, in *Evolution: The Grand Experiment*, examines the probability of the evolution of the whale from hyenas via mutations. He did some basic calculations of the number of changes caused by mutations that it would take to change the DNA code of a hyena type creature into the DNA for a whale. His guesstimate on the possibility of these changes occurring through mutations was about 1 out of 364 followed by 1,625 zeros. He writes: *"In other words, the chance of a hyena becoming a whale may be less likely than the chance of winning the national Powerball Lottery every year in a row for 200 straight years.*

Or the odds may be less likely than throwing 2,000 dice (at once) and all coming up as a '3'."[44]

The belief in lucky mutations as the engine of evolution is in even deeper trouble than the infinitesimal probabilities indicate. In life, as in science, a mistake may occasionally lead to some positive development, but no one in their right mind would suggest that mistakes are the powerhouse for the development of technology or scientific knowledge. The more mistakes there are in the process of creating any machine the more likely that the machine will fail. Yet evolutionists propose that random genetic mistakes are the mechanism through which our cells, our bodies, and our brains just happen to have been transformed from a single cell.

Random mutations are not the engine of evolutionary transformation but rather the end of any hope of real evolutionary development. Most evolutionists clearly acknowledge that most mutations are negative or harmful to the organism. Even if there were a few genetic mutations that may arguably increase the genetic pool and give some kind of advantage to an organism, there wouldn't be nearly enough to account for the number of positive mutations needed to transform one kind of creature into another kind. Since most mutations are either neutral or harmful to an organism, mutations will, given enough time, lead not to the evolution of a new type of creature, but rather to their extinction. Mistakes don't create a Boeing 747, but they may lead to its crash.

Dr. John C. Sanford is a plant geneticist who has published over eighty articles in peer reviewed scientific journals and has thirty patents on different scientific related inventions including the "gene gun". In 2005 he wrote *Genetic Entropy & the Mystery of the Genome* in which he states that, due to accumulated mutations, the human genome is deteriorating so quickly that it could not have evolved over hundreds of thousands or millions of years.[45]

As a consequence of recessive accumulated mutations the human genome is quickly decaying and not evolving in a positive direction. Although Dr. Sanford was an atheist at one time, he, like many other scientists when confronted with the facts of experimental science, turned his thoughts in GOD'S direction. He became a theistic evolutionist for a while and then a committed Genesis creationist.

> According to his own words, he did not fully reject Darwinian evolution until the year 2000... An analogy Sanford uses to illustrate alleged evidence of design is that of a car versus a junkyard: "A car is complex, but so is a junkyard. However, a car is complex in a way that is very specific — which is why it works. It requires a host of very intelligent engineers to specify its complexity, so it is a functional whole."[46]

Dr. Jonathon Sarfati, in commenting on Dr. Sanford's research, also mentions that the deterioration in the human genome fairly closely matches the shortening of the life spans of the Biblical patriarchs revealed in genealogies in Genesis 5 and 11.[47]

It is common for fairy tales to begin with "a long time ago in a far away land". This is also the technique of evolutionists to justify their amazing tales of how single cells mutated into princes through the magical kiss of time. They seek to inspire people to imagine a far away land, different from our world today, where magical mutations slowly but surely create incredible creatures with amazing abilities. The experimental reality is that mutations are no magic; they are mistakes that inevitably lead to extinction.

A FLOOD OF FOSSILS

Dr. Carl Werner devotes eleven of the twenty chapters in *Evolution: The Grand Experiment* to providing an amazingly comprehensive overview of the fossil record. There are a lot of fossils in our world but none of them come stamped with a date. The fossils are real but the dates assigned to them are only interpretations based on certain evolutionary assumptions. After visiting many museums around the world Dr. Werner estimates that there are presently over 200,000,000 fossils in various museums on our planet. This includes creatures big and tall, soft and small: dinosaurs, plants, worms, insects, fish eggs, turtles, jellyfish, bats, birds, and bees, to mention a few. His book is loaded with many photos of some of these fossils. Interestingly, the fossil record contains most of the kinds of creatures found in our world today.

Dr. Lance Grande, the curator of the Department of Geology, Chicago Field Museum, stated that the Green River formation in Wyoming contains billions of fish fossils that have yet to be excavated.[48] In the Karoo formation in Africa scientists have estimated that there are fossil remains of hundreds of billions of animals although, obviously, these are wild guesstimates, as no one has counted them. Most continents are covered with layers of sedimentary rock about a mile deep and there are an incredible number of fossils in many of these layers.

Despite the brazen attempt by evolutionists to appropriate the fossil record as a testimony to evolution, the truth is that the amazing number of fossils point directly to the reality of the incredible global flood testified to in Genesis 6. According to the timeline that GOD has given us in the Bible, the flood event took place approximately sixteen or seventeen hundred years after creation, about seven hundred years after Adam and Eve had passed away. This was plenty of time for billions of creatures to have populated the earth. In the description given of the world in the last hundred years before the flood we are told that there was a human population explosion *(Genesis 6:1)*. The number of people living in the world at the time of Noah isn't revealed, but based on population statistics guesstimates it could easily have been in the hundreds of millions or even more.

What happened to the billions of creatures that died in the flood? All over the earth some of their remains are found as fossils in sedimentary rock layers that were laid down by the waters of the flood. You couldn't have any clearer testimony to the reality of the flood than the fossil record. Billions of fossils are found in the geological layers, and each fossil is a testimony to some cataclysmic event that took place in the past.

Evolutionists maintain that the layers were laid down slowly over millions of years through erosion and sedimentation. However, it is a scientific fact that fossils do not form through a slow process of erosion and sedimentation. If a creature isn't buried quickly, it rots and returns to the elements. This is observational science and the common experience of humans. We all know what happens to most creatures when they die, including people: our physical bodies return to the physical elements, or *"to the dust"*, as Scripture teaches. A creature can be fossilized only through some cataclysmic event. The huge number of fossils in the geological record reveals that the geological layers themselves were largely produced by cataclysmic events.

Dr. Carl Wieland, director of Creation Ministries International, in his booklet *Stones and Bones,* has some interesting photos of an ichthyosaur that was fossilized in the process of giving birth, a fish in the process of eating another fish, and of trees that are fossilized standing straight up.[49] It doesn't take a tree very long to rot. It can only be fossilized if it is covered quickly with many meters of mud or dirt. Although a small percent of fossils undoubtedly formed from smaller localized floods or landslides, most of the hundreds of billions of them are a clear testimony to the unparalleled phenomenon of the global flood. During the flood, water and mud swiftly buried billions upon billions of creatures, ending their life and preserving the forms of their bodies through fossilization.

Recently, a fossil of a mega-dinosaur called Futalognkosaurus (meaning "giant chief lizard) was found in Argentina. It was estimated to be between thirty-two and thirty-four meters (105-111 feet) in length and about fourteen meters tall (forty-three feet), having a neck that was about nineteen meters long (fifty-six feet), and a spinal column that may have weighed up to nine tons.[50] Some have called it Behemoth referring to the enormous land creature that GOD describes in Job: *"Look at the behemoth, which I made along with you and which feeds on grass like an ox. What strength he has in his loins, what power in the muscles of his belly! His tail sways like a cedar; the sinews of his thighs are close-knit. His bones are tubes of bronze, his limbs like rods of iron. (Job 40:15-18)* It would take an immense cataclysm, like a global flood, to fossilize a huge creature like Futalognkosaurus.

WHERE ARE THE APE-PEOPLE AND OTHER MUTATING CREATURES?

Earlier in this chapter we focused on the truth that life comes only from life, and one of the basic biological laws of life itself is that cats come from cats, dogs from dogs, finches from finches, monkeys from monkeys, and people from people. This is a scientific verifiable fact that has never experimentally or observationally been disproven. Life in our world reflects the Biblical creationist teaching of Genesis 1 where GOD reveals that He created each kind as genetically unique, so that there is no natural intermingling between them. Due to the genetic program for each kind, there is much variety within each kind, but not any changes indicating one type of creature is evolving into another.

If evolution were taking place around us we wouldn't just have minor changes within species, but thousands of intermediary creatures in the process of evolving into new kinds. Apes managed to survive, but all of the more highly evolved ape-people have all disappeared. There isn't a creature on earth that is a clear example of an intermediate kind. If the theory of evolution was more than just the overwrought imaginations of some unbelieving scientists we would be surrounded by many intermediary kinds of life in the various stages of mutating from one kind to another. Clearly identified transitional kinds don't exist, either in the fossil record or in the biosphere around us.

We see apes and people but where are all the ape-people? If evolution were true there must have been at least thousands of intermediate species or kinds between apes and humans. Where are all of the missing links between apes and people? An individual needs an incredibly great imagination to believe that apes survived for millions of years, but all the more intelligent ape-people have all magically disappeared because they couldn't handle the changing environment or the competition from the apes.

If evolution had taken place, the fossil record should reflect the reality of many intermediate kinds of creatures. Even Darwin acknowledged that the fossil record didn't match his theory. He wrote:

> Geology assuredly does not reveal any such finely-graduated organic chain; and this, perhaps, is the most obvious and serious objection which can be urged against the theory. The explanation lies, as I

believe, in the extreme imperfection of the geological record."[51]

Dr. Andrew Knoll, a professor of biology at Harvard University, stated in a 1998 interview with Dr. Carl Werner:

"Darwin devotes two chapters of The Origin [of Species] to the fossil record. And you might think that's because Darwin, like most of his intellectual descendents, would have seen the fossil record as the confirmation of his theory. That you could really see, directly document, the evolution of life from the Cambrian to the present. But, in fact, when you read The Origin [of Species], it turns out that Darwin's two chapters are a carefully worded apology in which he argues that natural selection is correct despite the fact that the fossils don't particularly support it."[52]

Darwin believed, he didn't know, but he believed, the reason the fossil record didn't demonstrate his story of evolution was simply because the geological record was imperfect, and that with time many, if not most, of the missing links between kinds of creatures would be found. One hundred fifty years later, after billions upon billions of fossils have been found all over the earth, providing a reasonably clear picture of the creatures which existed in the past, the links are still missing. One explanation that many evolutionists give for the absence of transitional fossils is that it is very unusual for a fossil to form. However, despite how difficult it is for fossils to form there are billions of them all over the face of the earth providing a good representation of the earth's past inhabitants. With geologists digging all over the earth in their search for oil, gas and minerals, and tens of thousands of palaeontologists sifting the earth for missing links for over a century, evolutionists have only a few pieces of disputed fossils that they imagine are missing links.

One of the facts of the fossil record that evolutionists don't trumpet is that most fossils are easily recognizable as different species of the same kinds of life that we have in our world today. The fossil record is largely a record of stasis involving only minor changes in species. Most kinds of creatures that exist in our world today including birds, fish, insects, land animals and even plants are also present in the fossil record. This fits in well with the reality of the flood, but is contrary to what should exist if evolution were true.

Undeniably, there are some species and even kinds of creatures, such as dinosaurs, that exist in the fossil record but are not found alive anywhere in our world today, that we know of. The fossil record clearly reveals the extinction of certain species and even kinds of creatures, but what it does not reveal is the evolutionary tree of millions of transitional kinds. Note that the extinction from the biosphere of a few plants and animals, like dinosaurs, since the time of the flood is in complete harmony with how the eco-systems would have been transformed as a result of the flood, and there would also be increasing decay in all life as a result of humanity's fall into darkness (More on this later in chapter 5).

MIRACLE BLOOD OR A MIRAGE OF TIME?

The evolutionary story says that all of the dinosaurs were wiped off the face of the earth about sixty-five million years ago. Recently they excavated some un-fossilized dinosaur bones. Dr. Mary Schweitzer, an evolutionary palaeontologist, has even identified soft dinosaur tissue, including parts of blood vessels and cells, in some of these un-fossilized dinosaur bones.[53] Incredibly, many evolutionists now believe that soft dinosaur tissue can remain intact without fossilization for seventy million years by some unknown miraculous process. Even though all practical experimental science demonstrates that soft tissue deteriorates fairly rapidly after death, they have to believe that this tissue miraculously survived for seventy million years. They cannot face the obvious reality that perhaps these dinosaurs are only a few thousand years old.

What would be the significance if their dating methods for dinosaurs were so blatantly inaccurate? It would invalidate all of their geological time frames and demolish their illusion of a materialistic evolutionary world. If, in order for evolution to be true, they need to believe that soft dinosaur tissue and parts of blood cells can survive for tens of millions of years, defying experimental science, then their evolutionary faith is certainly up to the challenge. This is a case, not of miraculous blood, but of evolutionists having a mirage of time.

At one time dinosaurs were used as propaganda tools to try to silence those who believe in the creation account in Genesis. Now the tables have turned and dinosaur blood is providing clear testimony to a fairly young earth.

The only miracle blood I know of is that of JESUS CHRIST. CHRIST died for the sin of the world, your sin and my sin, so that by faith in Him we could have the knowledge of the forgiveness of our sin and the glorious hope of eternal life. There is amazing saving power in the blood of CHRIST for everyone who repents and turns to Him. He is the real Mr. Clean who can clean your whole life and everything in it.

> *"To him who loves us and has freed us from our sins by his blood, and has made us to be a kingdom and priests to serve his God and Father—to him be glory and power forever and ever! Amen." (Revelation 1:6)*

THE POPULATION BOMB EXPLODES EVOLUTIONISM

Our planet's population has soared from a few people to over six and a half billion during the time of recorded history, a mere six thousand years. After two world wars, hundreds of smaller ones, extermination programs for tens of millions of their own people carried out by the great human tyrants of the 20th century, the abortion of over fifty million unborn children each year for the past half century, malaria, AIDS, TB, cancer, and other plagues and diseases, the population of our world has mushroomed from one and a half billion to over six billion in the past century alone. The facts of life explain why populations increase despite all of these other factors. The only nations that haven't recently increased in population as a result of births are the western nations with the highest selfish material standard of living.

Most evolutionists believe that modern man, Homo sapiens, evolved into our present state somewhere between 200,000 to 50,000 years ago. If the human race has really existed for 50,000 years, even with wars, plagues, and violence, there would have been hundreds of millions, if not billions of people within the first 10,000 years, and who knows how many since then. So what happened to all of these people? Why was the population only a couple hundred million people at the time of CHRIST two thousand years ago?

The story that evolutionists have concocted to explain why human populations remained so small during these 45,000 or more imagined years before recorded human history is that the people then lived like animals, and that most of them died for lack of food or from various diseases in infancy or childhood. Does this story make any sense in light of the reality that the population in modern times has increased most dramatically in the countries with the worst health care, the least education, the highest birth and childhood mortality rates, and the greatest poverty? If lack of food had been a problem, many of them would have been smart enough to have migrated to other places where there was food, just like the creatures of the wild do, and as people have always done during the time of written history. What about curiosity? Wouldn't some of them have been curious enough to explore for greener pastures? With their intelligence they would have learned that many plants grow from seeds and that they could plant those seeds and harvest their own food.

The evolutionary story that prehistoric people lived like animals, communicated through grunts, and lived in caves or forests for tens of thousands of years without developing language, building homes, learning to plant seeds, developing agriculture, raising domesticated animals, caring for their sick, or even looking after their children doesn't even fit with their evolutionary slogan of "survival of the fittest". By this slogan, the smartest, toughest, best breeders, with the best immune systems, would have survived the longest in each generation, making the subsequent generations smarter, tougher, better breeders with stronger immune systems. This is doctrinaire Darwinism. Why then, do evolutionists believe that all these people lived as weak, powerless, low-browed "doofuses" for tens of thousands of years? You have to wonder how it was possible for humans to have evolved brains that could accomplish so much if they never used them even to look after their own basic needs. Is it logical to believe that people magically evolved brains they never used until six thousand years ago?

There is no hard evidence for any of their speculations regarding these 50,000 years of supposed human existence. These conjectures are contrary to the record of history and contemporary population growth statistics. According to the facts of life, without any national health care programs delivering abortion, euthanasia, condoms, the pill, and sterilization services, populations would increase rapidly. Increasing populations would have resulted in people migrating and forming civilizations in every part of the world. This is exactly what has happened during the past few millennia, as GOD has accurately revealed in Genesis.

The population bomb is not only applicable to human history but to all forms of life created with an innate reproductive desire. Where are the trillions of miraculous bones from the dinosaurs that supposedly roamed the earth two hundred and thirty million to sixty million years ago? How many dinosaurs would there have been during this imaginary one hundred and seventy million year period? Let's do some basic math. The scientific estimate right now is that there were over five hundred different known species of dinosaurs during the so-called dinosaur era. Say that there were only about ten thousand individual dinosaurs for one percent of the species of dinosaur for every hundred years of one hundred and seventy million years (a very, very, very low estimate based on animal instincts). Here is the math: 10,000 dinosaurs (per 100 years) X 5 species of dinosaurs (1% of total known species) X 10 (1,000 years) X 1000 (1 million years) X 170 million years equals 85 billion dinosaurs. Remember this is an exceptionally low estimate; a more realistic estimate would probably have been at least thirty to fifty times as many during this imaginary dinosaur age. Where are all the trillions of extremely miraculous dinosaur bones and flesh that can be preserved un-fossilized for tens of millions of years? If this were true, our planet would be littered with dinosaur bones.

Then there are the rabbits. Can you imagine how many rabbits would have existed on our planet if they had been around for the thirty million years that evolutionists proclaim? How about rats?

Whether it is human or animal populations, the facts of life don't lie: life in our world can only be a few thousand years old.

TIME TO TURN THE PAGE

The mythologies of cosmic and biological evolution are contrary to logic, the principles of math, and the scientific facts of life, yet they are believed and taught as the generally accepted scientific theory of origins in our world today. What does all of this mean? I believe it means we need to take a closer look at the legitimate place and role of science in the human enterprise. It is time to quit worshipping science and to acknowledge its great limitations. It is time to turn the page to a more realistic and practical understanding of the place and role of science in human history.

CHAPTER 3

Scientists Aren't GOD and Science is no Savior

"Who is this that obscures my plans with words without knowledge? Brace yourself like a man; I will question you, and you shall answer me. Where were you when I laid the earth's foundation? Tell me, if you understand. Who marked off its dimensions? Surely you know!...What is the way to the abode of light? And where does darkness reside? ... Can you bind the beautiful Pleiades? Can you loose the cords of Orion? Can you bring out the constellations in their seasons, or lead out the Bear with its cubs? Do you know the laws of the heavens?... Will the one who contends with the Almighty correct Him? Let him who accuses GOD answer Him!" (Job 38:2-5,19, 31-33; 40:2)

There is a mountain of propaganda proclaiming that evolution is a "scientific fact" and the only scientific theory that has any credibility. As demonstrated in the last chapter, nothing could be further from the truth. Nevertheless, many feel caught between a rock and a hard place, between what they believe or at least were taught about GOD, and what they are being taught by many scientists regarding our origins. What many do not seem to have understood is that scientists are not, never have been, and never will be GOD. Secondly, science operates with a very narrow materialistic view of life and cannot present us with the total picture of the dynamic complexity of life involving the mind and spirit as well as the body.

HUMANISM AND THE WORSHIP OF SCIENCE

During my lifetime, science has increasingly become an idol to many, and scientists have been worshipped or venerated as super-humans or even infallible gods. The combination of the religion of secular humanism and science fiction mythologies has persuaded many to believe that science is humanity's only hope for a utopian future. One of the clearest expressions of this over the past century is found in the statements of secular humanists who believe that GOD is just a myth from the past, and that science is the only way to an enlightened and exhilarating future. Here are just a few excerpts from the Humanist Manifestos 1, 2, and 3 that highlight this way of thinking.

Humanist Manifesto 1 1933

> Today man's larger understanding of the universe, his scientific achievements, and deeper appreciation of brotherhood, have created a situation which requires a new statement of the means and purposes of religion. Such a vital, fearless, and frank religion capable of furnishing adequate social goals and personal satisfactions may appear to many people as a complete break with the past. While this age does owe a vast debt to the traditional religions,

it is none the less obvious that any religion that can hope to be a synthesizing and dynamic force for today must be shaped for the needs of this age. To establish such a religion is a major necessity of the present. It is a responsibility which rests upon this generation. We therefore affirm the following:

FIRST: Religious humanists regard the universe as self-existing and not created.

SECOND: Humanism believes that man is a part of nature and that he has emerged as a result of a continuous process.

THIRD: Holding an organic view of life, humanists find that the traditional dualism of mind and body must be rejected.

FOURTH: Humanism recognizes that man's religious culture and civilization, as clearly depicted by anthropology and history, are the product of a gradual development due to his interaction with his natural environment and with his social heritage. The individual born into a particular culture is largely molded by that culture.

FIFTH: Humanism asserts that the nature of the universe depicted by modern science makes unacceptable any supernatural or cosmic guarantees of human values. Obviously humanism does not deny the possibility of realities as yet undiscovered, but it does insist that the way to determine the existence and value of any and all realities is by means of intelligent inquiry and by the assessment of their relations to human needs. Religion must formulate its hopes and plans in the light of the scientific spirit and method.

SIXTH: We are convinced that the time has passed for theism, deism, modernism, and the several varieties of "new thought".[54]

Humanist Manifesto 2 1973

The next century can be and should be the humanistic century. Dramatic scientific, technological, and ever-accelerating social and political changes crowd our awareness. We have virtually conquered the planet, explored the moon, overcome the natural limits of travel and communication; we stand at the dawn of a new age, ready to move farther into space and perhaps inhabit other planets. Using technology wisely, we can control our environment, conquer poverty, markedly reduce disease, extend our life-span, significantly modify our behavior, alter the course of human evolution and cultural development, unlock vast new powers, and provide humankind with unparalleled opportunity for achieving an abundant and meaningful life.[55]

Humanist Manifesto 3 2003

Humans are an integral part of nature, the result of unguided evolutionary change. Humanists recognize nature as self-existing. We accept our life as all and enough, distinguishing things as they are from things as we might wish or imagine them to be. We welcome the challenges of the future, and are drawn to and undaunted by the yet to be known.[56]

The precise purpose of humanism is to create a new religion based on the belief in the non-existence of GOD, the worship of science and the faith that humanity can solve all problems and be its own savior. Humanism is a false religion that worships human knowledge and presents a futile hope to those who choose to live without GOD. Religious humanists reject GOD as Creator, believe in evolution, which is foundational to their other beliefs, and reject the existence of mind and soul. They believe that all religions, including Christianity, are merely the evolutionary product of their heritage and culture. The devout humanist believes that only modern science, in partnership with humanism, can bring hope to our world. Since GOD isn't real to them, humans usurp GOD'S place as the sole creative intelligence and power that can impact the world, and eventually the universe.

Unfortunately, secular humanists are not alone in worshipping science and scientists. Many church leaders, theologians, and preachers have also placed science on the throne and believe only what science tells them they can believe. In the last few years, some church leaders have set aside a special Sunday to praise Saint Darwin for setting people free from believing that the GOD of the Bible is their Creator.[57] In this scientific culture that permeates society, people's practical beliefs are often determined by the materialistic doctrines of science and not by a true knowledge of and relationship with GOD through His SON, SPIRIT and Word. Undoubtedly, for many people, science has become their idol or god, and Darwin, their patron saint or holy father.

CHILDREN IN THE LITTLE LEAGUES

"Then GOD said, "Let Us make mankind in Our image, in Our likeness, so that they may rule over the fish in the sea and the birds in the sky, over the livestock and all the wild animals, and over all the

creatures that move along the ground. So GOD created mankind in His own image, in the image of GOD He created them; male and female He created them. GOD blessed them and said to them, "Be fruitful and increase in number; fill the earth and subdue it. Rule over the fish in the sea and the birds in the sky and over every living creature that moves on the ground." (Genesis 1:26-28)

We are not the eternal, almighty, Holy GOD, Creator of the heavens and the earth, but we were created to be His children with incredible abilities that GOD has given us. As people, created in the image of GOD, we have been given minds that are able to create and produce incredibly intricate, beautiful, and bountiful works of art, literature, industry, agriculture, athletics, commerce, government, and science. The Christian understanding has never been to denigrate the abilities that GOD has invested in humanity. Even when these abilities were used for evil purposes, in the Tower of Babel event, the LORD said: *"If as one people speaking the same language they have begun to do this, then nothing they plan to do will be impossible for them." (Genesis 11:6)* Using our minds and the laws GOD has established in our universe, there is no end to what humanity could do, although, as a result of our rebellious nature, much of it would be destructive of human life and society.

The ability to learn, analyse, invent, create and minister to each other is a wonderful gift GOD has given us for our benefit and the blessing of all creatures. Although we are not in GOD'S league when it comes to creating things, even in the little leagues we can develop some astounding contraptions: computers, cell phones, and jets. Who knows what new discoveries and technologies are around the corner? One hundred years ago many of the technologies we take for granted today would all have been viewed as miraculous.

Of course now we recognize that they are not miracles, but very creative ways of using the laws GOD has established for the purposes of the human race. This is the proper exercise of legitimate science and we should be thankful to GOD and scientists for many of these helpful inventions.

All people, not only those engaged in scientific endeavors, can be a little bit like GOD as we use our GOD-given gifts and abilities in positive ways to love and serve GOD and our neighbors. A clear stream of JESUS' teaching and of the teaching of the HOLY SPIRIT in the Bible is that, as GOD'S children, we are to be holy as GOD is Holy. We are to be loving and forgiving like GOD, like CHRIST. We are to be "little Christs" in our world, like JESUS in our ministry to others. Nevertheless, we need to always remember that we are in the little leagues and that GOD is the one that we worship and seek to conform ourselves to. He is the only One worthy of our ultimate adoration and praise. **JESUS said:** ***"You shall worship the LORD your GOD, and Him only shall you serve!" (Matthew 4:10)***

CREATION SCIENTISTS OF THE PAST

Some scholars assert that science was stillborn in most cultures and developed in the western world largely as a result of the Christian understanding of our universe and life, as an orderly creation of GOD. There is no question that the history of science is loaded with many Biblical creationists. On the Creation-Evolution Headlines website David F. Coppedge has a listing of creation scientists between 1000 – 2000 AD:

<u>THE EARLY CHRISTIAN ROOTS OF MODERN SCIENCE</u>

- The Medieval Philosophers: Hugh, Ockham, Oresme
- Robert Grosseteste – Nature is knowable

- Roger Bacon – Experiment is the key
- Leonardo da Vinci – Master of all trades
- Sir Francis Bacon – Pathfinder to truth relies on God's word
- Johannes Kepler – Thinking God's thoughts after Him
- William Harvey – Surgeon to King James reveals secrets of the circulatory system

SCIENCE TAKES OFF IN ALL DIRECTIONS

- Blaise Pascal – The short-lived genius, passionate for Christ Jesus
- Robert Boyle – Leading experimenter leaves a legacy to fight skepticism
- Sir Isaac Newton – Left the universe a different place, in answer to prayer
- Antony van Leeuwenhoek – The shop merchant in awe of God's tiny creatures
- Carolus Linnaeus – Organizer of the Genesis kinds
- William Herschel – An undevout astronomer must be mad
- John Herschel – All scientific findings confirm Scripture
- Samuel F. B. Morse – What hath God wrought!

"NATURAL PHILOSOPHY" REACHES ITS ZENITH

- Michael Faraday – World's greatest experimental physicist, a humble, Bible-believing Christian
- Charles Babbage – Father of the computer defends the Scripture
- James Prescott Joule – Father of thermodynamics does science to ponder God's wisdom

- Lord Kelvin – Eminent physicist/professor takes on Darwin and his bulldog
- James Clerk Maxwell – Christian creation scientist par excellence
- Great Christian Mathematicians: John Napier, Leonhard Euler, Bernhard Riemann
- Honorable Mentions in Physical Science: Copernicus, Brahe, Flamsteed, Davy, Dalton, Henry, Fleming

SHINING THROUGH MATERIALISTIC DARKNESS

- Gregor Mendel – The monk whose gene laws Darwinists had to obey
- Louis Pasteur – World's greatest biologist opposes evolutionism
- Joseph Lister – Compassionate Quaker saves millions of lives
- The Anti-Evolutionists: Not just Bible-believers opposed Darwin's ideas
- Honorable Mentions in Life Sciences: Ray, Hooke, Bell, Simpson, Fabre
- Henrietta Swan Leavitt – The gentle Christian lady PhD who measured the universe
- George Washington Carver – Obedience to the Genesis mandate saves the South
- Wernher von Braun – World's greatest rocket scientist defends Genesis
- James Irwin – The Apollo astronaut who took the Bible to the moon

THE RESURRECTION OF CREATION SCIENCE

- A. E. Wilder-Smith – Triple-PhD chemist pioneers intelligent design reasoning

- Raymond V. Damadian –Creationist revolutionizes diagnostic medicine
- Henry M. Morris – Father of the modern scientific creationism movement
- Duane Gish: The man the Darwinist debaters feared most
- Stephen A. Austin: Bringing Genesis back to the real world
- Richard D. Lumsden – Scientism can't save the scientist's soul[58]

This very incomplete list of Christian believers involved in the history of science is not to suggest that the contributions of others were not also extensive. It does highlight the fact, however, that the scientific endeavour of seeking to understand this world of ours is in harmony with a Christian understanding of life and indeed flourished because of it.

The material facts of our universe and life are the same regardless of who uncovers them. Those scientists who are not believers in CHRIST have nevertheless been created by GOD with incredible abilities. It is not an issue of Christian scientists being more intelligent or capable than secular scientists or vice-versa. The critical issue is the use of scientific abilities in ways which are in harmony with GOD'S good will and as a benefit, not a detriment, to humanity and all creatures.

BEWARE THE BOASTING OF THE WHITE COATS

"This is what the LORD says—Israel's King and Redeemer, the LORD Almighty: I am the first and I am the last; apart from Me there is no god. Who then is like Me? Let him proclaim it. Let him declare and lay out before Me what has happened since I established My ancient people, and what is yet to

come—yes, let him foretell what will come. Do not tremble, do not be afraid. Did I not proclaim this and foretell it long ago? You are my witnesses. Is there any GOD besides Me? No, there is no other Rock; I know not one." (Isaiah 44:6-8)

Only GOD knows the totality of history and what the future will bring, yet many scientists confidently and continuously assert that they know the past and can predict aspects of the future. Understandably, those who don't know the Living GOD, the Eternal I AM, cling to a variety of idols, including science and scientists. These are all false and deceptive hopes. Scientists cannot save even themselves, let alone save all of humanity.

None of us is GOD. When individuals begin to believe that they are more intelligent, more righteous, or more powerful than GOD, they have been deceived and their thinking becomes corrupted. The sixteenth century reformer, Martin Luther, wrote that apart from GOD'S Word and GOD'S SPIRIT, reason becomes the "whore of the devil". Intellectual ability, like the love of money, will often lead to a great deal of foolish arrogance and the temptation to use this endowment for selfish evil goals. As a result of sinful arrogance toward GOD and others, much of what is done in every area of life, including science, is contrary to the good will and purposes of the eternal I AM. The scientific accomplishments of Nazi Germany are an obvious illustration of this truth. Scientific knowledge, apart from true knowledge of the living GOD, will often be corrupted and used for purposes contrary to GOD'S good will and detrimental to the welfare of humanity.

The tree in the Garden of Eden was the tree of the knowledge of good and evil. The knowledge gained by science is neutral, but as a result of humanity's fallen nature it is often used for nefarious purposes. It is foreseeable that, as a result of arrogance and self-centeredness, much of the knowledge appropriated by science may yet prove to be more destructive than beneficial to humanity.

Beware of the boasting of the "white coats". Despite all of the hype and crowing, science and scientists in every age are very limited in what they can do. Even if they choose to use their knowledge in positive ways, the most they can do is to extend life a few months or years and make some dynamics of living easier. This is good and helpful, but also limited.

Interestingly, even with all of our scientific knowledge and incredible technologies today, the average lifespan in the developed world is still between seventy and eighty years, just as it was for the Hebrews in the days of King David three thousand years ago: *"**The length of our days is seventy years—or eighty, if we have the strength." (Psalm 90:10)***

I recall hearing in the sixties and seventies that through science most diseases would be conquered within twenty to thirty years. That time has long passed, and, instead of victory over most diseases, we are experiencing the advent of strange new diseases and the return of worldwide plagues. This is certainly a fulfillment of JESUS' prophecies regarding the last days before His return (Luke 21:10,11).

Science cannot save anyone for very long. Scientists cannot create life and they cannot extend it for more than a few years at best. We can be thankful for the help that scientists are able to give us for the time we have in this world. They are our neighbors who have been gifted by GOD, and many have been trained and use their gifts for the benefit of people. Thank them, praise GOD for them, but don't worship them.

For secularists to place their faith and trust in science to save their lives and our planet is predictable, since, in their narrow materialistic view, they have nothing else to trust in. What is almost incomprehensible is that those who claim to know GOD and CHRIST would be willing to prostrate their minds in worship and praise of science and to place more faith in scientists than in GOD. There are those who will believe every wild imaginary idea that scientists offer up, particularly evolution, but steadfastly refuse to believe what GOD says concerning what He has done in the past and promises to do in the future. They believe that scientists can do miracles but that GOD cannot or will not do anything miraculous. Scientists are now able to create diamonds within days, but continue to insist that it took millions of years to create the diamonds we mine from the earth. Via microchips we can talk to people on the other side of the world, but many refuse to believe that GOD can or would talk to anyone. For many people, what scientists say automatically overrules anything that GOD has said through His HOLY SPIRIT. Christians who foolishly disregard the clear revelation of GOD in His Word for the boastings of science are exchanging their birthright, as Esau did, for a mess of pottage. If they continue down Esau's path, they will lose their inheritance and make their home with the world.

FAITH IN THE SCIENTIFIC EXPERTS

The most common argument from those who believe the evolutionary story is the appeal to authority. Most people who believe in evolution do so because they have total trust in the scientific experts.

Dr. Russell Humphreys, a former researcher in nuclear physics, geophysics, theoretical atomic and nuclear physics at Sandia National Laboratories, New Mexico, has a long list of scientific achievements and inventions to his name. He was once an ardent evolutionist and atheist, but is now a convinced creationist. In a recent article entitled: *"Why most scientists believe the world is old: Beliefs Foster Further*

Beliefs", on the Creation Ministries International website (April 1, 2010) he wrote:

> "There are many categories of evidence for the age of the earth and the cosmos that indicate they are much younger than is generally asserted today. There is a little-known irony in the controversy between creationists and evolutionists about the age of the world. The majority of scientists— the evolutionists—rely on a *minority* of the relevant data. Yet a minority of scientists—the creationists— use the *majority* of the relevant data. Adding to the irony is the public's wrong impression that it is the other way around. Therefore, many ask: *"If the evidence is so strongly for a young earth, why do most scientists believe otherwise?"* The answer is simple: <u>Most scientists believe the earth is old because they believe most *other* scientists believe the earth is old!</u>
>
> Going round in circles
>
> They trust in what's called 'circular reasoning', not data. I once encountered such a clear example of this misplaced trust, that I made detailed notes immediately. It happened when I spoke with a young (in his early thirties, career-ambitious, and upwardly mobile) geochemist at Sandia National Laboratories, where I then worked as a physicist. I presented him with one piece of evidence for a young world, the rapid accumulation of sodium in the ocean. It was ideal, since much of geochemistry deals with chemicals in the ocean.
>
> I wanted to see how he explained possible ways for sodium to get out of the sea fast enough to balance the rapid input of sodium to the sea. Creationist geologist Steve Austin and I wanted the information in order to complete a scientific paper on the topic.

We went around and around the issue for an hour, but he finally admitted he knew of no way to remove sodium from the sea fast enough. That would mean the sea could not be billions of years old. Realizing that, he said, *"Since we know from other sciences that the ocean is billions of years old, such a removal process must exist."*

I questioned whether we 'know' that at all and started to mention some of the other evidence for a young world. He interrupted me, agreeing that he probably didn't know even one percent of such data, since the science journals he depended on had not pointed it out as being important. But he did not want to examine the evidence for himself, because, he said, *"People I trust don't accept creation!"*

Faith, not science

I asked him which people he was relying upon. His answer was, *"I trust Steven Jay Gould!"* (At that time Gould, a paleontologist, was still alive and considered the world's most prominent evolutionist.) Thus the geochemist revealed his main reason for thinking the earth is old: *"people I trust"* i.e., scientific authorities, had declared it. I was surprised that he didn't see the logical inconsistency of his own position. He trusted Gould and other authorities but ignored highly relevant data!

Perhaps the geochemist thought it so unlikely the earth is young that he wasn't going to waste time investigating the possibility himself. But if that were the case, then it shows another way the old-world myth perpetuates itself—by intellectual inertia.

I remember having similar attitudes when I was a grad student in physics, while I was still an evolutionist. I was wondering about a seeming inconsistency in biological evolutionism. But, I told myself, surely the experts know the answer, and I've got my dissertation research to do. I had no idea that (a) the experts had no answer for it, and (b) the implications were extremely important, affecting my entire worldview.

Before I became a Christian, I resisted evidence for a recent creation because of its spiritual implications. The geochemist might also be resisting such implications, and was merely using scientific authority as a convenient excuse.

The bottom line

Many scientists are not the independent seekers of truth the public imagines, so the public should not trust them blindly. For a variety of reasons, scientists depend on other scientists to be correct, even when they themselves have some reason for doubt. Unfortunately, as most creationist scientists can tell you, the young geochemist's reaction is not at all exceptional. Many scientists, without serious questioning, trust the opinions of their own 'experts'. However, I'm happy to report that others, when presented with creationist data, have become very interested and have investigated it. Many have become creationists that way, as I did. [59]

It is written: "It is better to take refuge in the LORD than to trust in man. It is better to take refuge in the LORD than to trust in princes." (Psalm 118:8) Notice that GOD repeats the message, just in case we didn't catch it the first time. Scientists are generally intelligent and we can respect their intelligence without revering them or believing everything they say. They do not know much of what has happened in the past. They were not there to observe the past and they cannot recreate it or perform repeatable verifiable experiments on what has already taken place. The past is beyond the reach of the tools of reliable experimental science. The most scientists can do is to speculate or make inferences based on present data. Their speculations are often just a reflection of their own particular biases rather than what actually transpired in the past. What an individual believes about the early history of man is largely based, not on actual data from the past, but on a particular faith, whether in scientists or GOD.

Many scientists believe that evolution is a scientific fact when in reality it is the conclusion of their materialistic presuppositions. The central assumption of evolutionists is that material is all that there is and therefore there is no GOD and no spiritual reality to life. However, there are many other presuppositions for evolution that are built on this materialistic foundation. In his book, *The Evolution of a Creationist,* Dr. Jobe Martin has a chapter called *"...And then Came Assumptions"* in which he quotes the evolutionist G.A Kerkut in listing a few of the major assumptions of those who hold to the evolutionary story. Dr. Martin wrote:

> These are the basic ideas an evolutionist "takes for granted" or "supposes" to be true. All of the "molecules-to-man science" is built upon these assumptions, but you rarely, if ever, see them listed in a high school or college textbook.[60]

Dr. Martin then quoted Dr. Kerkut:

> There are seven basic assumptions that are often not mentioned during discussions of evolution. Many evolutionists ignore the first six assumptions and only consider the seventh. The assumptions are as follows:

1. The first assumption is that non-living things gave rise to living material, i.e., spontaneous generation occurred.
2. The second assumption is that spontaneous generation occurred only once.
3. The third assumption is that viruses, bacteria, plants and animals are all related.
4. The fourth assumption is that protozoa (single-celled life forms) gave rise to metazoan (multiple-celled life forms).
5. The fifth assumption is that various invertebrate phyla are interrelated.
6. The sixth assumption is that the invertebrates gave rise to the vertebrates.
7. The seventh assumption is that within the vertebrates the fish gave rise to amphibian, the amphibian to reptiles and the reptiles to birds and mamamals."[61]

As Dr. Martin then points out, these assumptions include the whole of the evolutionary theory. The evolutionary story is made up of one assumption after another.

In Six Days: Why Fifty Scientists Choose to Believe in Creation Dr. John K. G. Kramer a research scientist with Agriculture and Agri-Food Canada, gave a few of his reasons why he believes in the Genesis creation account as an historical account as opposed to the evolutionary postulations:

> In my scientific career I have observed an interesting principle. Whenever little is known on a subject, or is different from the norm, more speculations arise as to its evolutionary development. Instead of admitting "we do not know," and working towards discovering the unknown, some evolutionary comments are usually made. On the other hand, the

more that is known about a certain subject, the more eagerness there is to describe it in detail, and classify such systems "irreducibly complex", as Michael Behe refers to it.

No one has ever demonstrated macroevolutionary changes on a molecular level, yet many people readily speculate evolutionary links between bacteria, plants, animals, and man. Are the gross structures not made up of individual cells with complex molecules? If macroevolution is unlikely at the molecular level, how can the whole be changed.[62]

As Dr. Kramer highlights, evolution is largely comprised of inventive speculations based on the mystery of the unknown rather than on a scientific explanation of what is known. It is a theory built out of a materialistic belief system, and not hard scientific facts.

HAPPY _____ BIRTHDAY TO THE EARTH

Let's take a closer look at the age of the earth mentioned earlier by Dr. Humphreys. As a general rule, most scientifically informed people who aren't Biblical creationists claim to know that the earth is about three and a half to four billion years old. So how do people know this? Do you know this? If you think you do it is probably because you have chosen to trust what some scientists have said or written. But how do they know? They certainly weren't there. They claim to know this "fact" by radiometric dating or other methods of measuring certain elements that are in the rocks and fossils to determine approximately how old a rock may be. Knowing that radioactive elements decay at a certain constant rate today, and claiming to know the approximate amount of these radioactive elements that was present in those rocks to begin with, they calculate approximately how much time has elapsed since the formation of those rocks and the fossils encased within them. Sounds logical, doesn't it? Or, does it?

Do they really know the amount of those elements that was present in the rocks or fossils at the time they were formed? No they do not. Their "knowledge" is based on assumptions, not data. Are they certain that nothing happened in the past that could have affected the rate of decay, either adding to or subtracting from the amount of radioactive materials that are present? No. Are they certain about what the earth was like when it was formed? No. Since scientists weren't present at the time the earth was formed or through its history, they have made certain assumptions based on models of what they believe the earth was like back then, and what its history has been since its beginning. That is how they do it: by making inferences or speculations based, not on actual scientific data from the past, but on their overall theory of what may have been past reality. Could their assumptions and models be wrong? Do scientists ever make mistakes?

Dr. Jonathon Sarfati, in his latest book, *The Greatest Hoax on Earth?*, has a chapter called: "Is the Earth Ancient". In it he highlights some of the huge problems with present dating methods applied to the age of the earth.[63] In the chapter after it, called "Young World Evidence", he shares recent data that points to a very young earth.[64] This new data includes the discovery of DNA, blood cells and soft tissue from extinct creatures, the decay in our earth's magnetic field, helium found in zircons, the concentration of salt in the oceans, and the presence of comets. (As an aside, unless you like to lose, it is best to stay away from playing chess against Dr. Sarfati. As a FIDE Master and former New Zealand chess champion, one of his pleasures is playing chess against up to twelve opponents at one time. The kicker is that he plays blindfolded and never loses. His latest book just as easily demolishes the evolutionary pretentions of Dr. Dawkins in *The Greatest Show on Earth*, which attempts to proclaim the glories of evolution.

This should help to dispel the myth that Creation Scientists do not think logically or understand evolution.)

Part of the model on which many evolutionists base their conjectures on the age of the earth is the "Big Bang" theory on the age and formation of the universe. Most scientists today insist that the universe is a little less than fourteen billion years old. Fifty years ago scientists "knew" that the universe was about twenty billion years old. What kind of math is that? How does fifty years equal a six billion year difference in the age of the universe? Apparently some of the assumptions changed based on new information gathered from their investigations of the universe. I wonder how old they will profess the universe to be fifty years from now. It all depends on what new data and theories they may have then.

Their "guesstimates" on the age of the universe are based on a little information combined with an enormous number of assumptions. Is it wise to make speculations about the past when you lack most of the information necessary to know the truth? It is like guessing on a picture of a puzzle when you only have a few of the pieces. Their guesses have more to do with their preconceived ideas than the actual information on the puzzle pieces. Some humility is in order.

Belief in the Big Bang theory is based on assumptions which are far from being proven, and which some physicists and cosmologists completely reject. The Big Bang Theory states that the whole universe came out of a theoretical mathematical point called a singularity. Despite the fact that no one really knows what a singularity is or what would have made it explode, many believe that space, time, light, gravity, matter, and the laws of nature have all emerged out of this bang. This includes mysterious realities like quantum mechanics or quantum physics that may override all of the other laws that govern our observable material universe.

Add to this confusion the dilemma that the Big Bang theory works theoretically only if our universe contained about ninety-five percent more mass and energy than present day science can account for. Since they don't know what or where this missing mass and energy is they have called it "dark matter" and "dark energy". What if research in the next fifty years discovers what this mysterious dark matter and energy is, where it is in our universe, and how it may affect radioactivity, the speed of light, and who knows what else? What then? What if it is still missing fifty years from now despite the billions of tax dollars they are putting into trying to find this mysterious stuff? What then? Will they toss out the Big Bang theory and substitute something else? It is invisible, but they "believe" it is there, because it has to be there for their theory to work.

Many scientists deny that the universe testifies to the existence of GOD, because you can't see GOD or put Him in a test tube. On the other hand, they believe that the universe proves the existence of singularities, of dark energy and dark matter even though they are all invisible, and may not even exist. They believe what they want to believe. The past, like the future, is beyond the scope of true experimental verifiable science and is easily molded to fit any philosophical assumptions that scientists may have.

Scientists aren't GOD. The more they stray away from reproducible experimental science focused on the present material realities in our universe and into all kinds of speculations about the past or the future, the less likely their "theories" will be true or helpful for humanity. The further they get from the present, the more their statements are mere speculations founded on their philosophical prejudices rather than on any known empirical facts. They reject the possibly that GOD'S Word is infallible, but strongly believe in their own infallibility and that of science.

WHEN TIME FLEW AND LIGHT STRETCHED

"In the beginning God created the heavens and the earth. Now the earth was formless and empty, darkness was over the surface of the deep, and the Spirit of God was hovering over the waters. And God said, "Let there be light," and there was light. God saw that the light was good, and He separated the light from the darkness. God called the light "day," and the darkness he called "night." And there was evening, and there was morning—the first day." (Genesis 1:1-5)

According to His testimony GOD created the heavens, the earth, water, space, light and time on the very first day. Before leaving the issue of the age of the universe and the world, it is important to recognize how the relationship between light, space, and time impacts our understanding of the age of the world and universe.

Based on observable, repeatable, experimental science it is fairly certain that light today travels through space at 300,000 kilometers or 186,000 miles per second. Thus light takes just over one second to be reflected to the earth from the moon, and about eight minutes to reach us from the sun. In one year light can travel approximately six trillion miles. This is known as a light year.

Physicists and astronomers have estimated the approximate distances to various stars and galaxies. According to their calculations many of these galaxies appear to be millions or billions of light years away from us. It seems logical therefore that those stars and galaxies are billions of years old. This has become a stumbling block, leading many to reject the teaching of the Bible regarding the age of the earth and universe. The age of the earth is approximately six thousand years based on the Bible's genealogical records

(Genesis 5, 11). Because of this, evolutionists, atheistic or theistic, ridicule and scorn those who hold faithfully to the Genesis account.

This is a good illustration of how materialistic science often jumps to conclusions without knowing all of the facts and how gullible theists merely accept the conclusions of materialist science without questioning or thinking it through. So what is wrong with the conclusion that the earth and universe are billions of years old? The problem lies in two areas, one is with respect to the speed of light, and the second is the nature of time.

All that scientists can conclude is that at the present time light travels at a specific speed, and as far as they know there is nothing in the universe that would affect the speed of light through space. This however, includes many presuppositions which cannot be empirically established, only assumed. Since there are many mysteries about and in the universe, including dark matter, dark energy, and quantum mechanics, it is impossible to conclude that none of these factors could impact the speed of light. There are simply far too many unknowns in the universe today to jump to premature materialistic conclusions.

Dr. Russ Humphreys, in his book, *Starlight and Time: Solving the Puzzle of Distant Starlight in a Young Universe*, presents a creation cosmology which could explain the problem of light from the stars reaching our planet in a shorter period of time. His cosmological theory is contrary to the generally accepted Big Bang theory and is thus generally discounted by those who toe the general "Big Bang" party line. His theory was partially inspired by Biblical passages in both Isaiah and Jeremiah where GOD declares that He stretched out the heavens. It is written: **"This is what the LORD says— Your Redeemer, Who formed you in the womb: I am the LORD, Who has made all things, Who alone stretched out the heavens, Who spread out the**

earth by Myself. (Isaiah 44:24) (Other verses expressing the stretching of the heavens: ***Isaiah 42:5; 45:12; 51;13; Jeremiah 10:12; 51:15)***

Dr. Humphreys suggests that, after the initial ex-nihilo (out of nothing) creation of the universe, GOD stretched out the heavens with our solar system near the center. This stretching of the heavens would also have included the stretching of light. Thus, the light from every segment of the cosmos would initially have been very close to the earth, but was then stretched out in the first few days of creation as the other galaxies were pulled away from the center. Consequently, there would be no issue of light from these distant galaxies having to traverse the astronomically huge distances to the earth. The light itself was initially fairly close to the earth and as these galaxies were stretched out from the earth, near the center of the universe, they would have left a trail of light behind them for the whole distance that they were stretched.[65]

The second aspect of Dr. Humphreys' cosmological theory deals with the nature of time. Time was created and began on the first day of creation. Time is a very mysterious entity as anyone who has lived for a few years in this world understands. Time is not a physical material reality, but a measurement idea, analogous to mathematics, created by GOD and placed into the human mind. It is an undeniable part of our life in this dimension, but not something that can be grasped or fully understood. Time, like math, is not a reality to any other creature as far as we can presently comprehend. Only humans are aware of the passage of time, of past, present and future tenses and of the concept of eternity. Humanity's life on earth is at a point of intersection between time and eternity. Time points us to the timeless GOD that we will focus on in the next few chapters.

One of the very peculiar properties of time is that it is not a constant in our universe; time changes. According to the theory of general relativity, in deep space time runs at a very different pace than on our planet. Therefore, the age of the universe depends on where in the universe you are. Even the satellites that circle our earth run in a slightly different time speed than our clocks on earth.

Dr. Humphreys theorizes that during the creation week, as the heavens were stretched out, time in deep space ran at an incredible pace, so that one creation day on earth could have been millions of years in deep space. Consequently, from our perspective on earth our universe is only about six thousand years old, but in deep space it is perhaps hundreds of millions or even billions of years old.[66]

This side of eternity it is impossible for any of us to wrap our minds around the reality that time is relative. Although Albert Einstein cannot take credit for creating time with its many mysteries, it was his theory of general relativity that opened these riddles of time for us. It is the same mystery that presents itself in science fiction movies or novels, where by going faster than the speed of light an individual is projected back in time. For all practical purposes of our life on earth, the issue does not really matter to us except in this one area: are we going to trust GOD'S clear historical account of the creation of the universe, world, and life, or, on the basis of many assumptions regarding the deep mysteries of light, space and time, are people going to reject the clear teaching of Genesis.

Is Dr. Humphreys' cosmological model true? I certainly don't know and no one else does either. He is a brilliant scientist who has come to know CHRIST and has a good comprehension of GOD'S Word. However, he is human and does not understand all of the mysteries of the universe. It is certainly possible that many parts of his model may be very accurate. Time may tell, but most of us will not know the answer this side of eternity.

What we do know is that Genesis gives us an historical account of the creation of the universe and world in six ordinary earth days approximately six thousand earth years ago. How GOD did it, including the mysterious realities of light, space, and time is beyond full human comprehension. It may be a worthwhile exercise for our minds to investigate these matters, but we all need to be humble and acknowledge that there is a great deal no one comprehends yet. Soon we will understand how it all fits together. Until then it is wise to place our greatest trust in GOD and not in human understanding. As GOD revealed to the prophet Habakkuk: *"For the revelation awaits an appointed time; it speaks of the end and will not prove false. Though it linger, wait for it; it will certainly come and will not delay. See, he is puffed up; his desires are not upright—but the righteous will live by his faith." (Habakkuk 2:3,4)*

SCIENTIFIC CONFUSION ON TESTABLE MATTERS

While the past certainly involves a great deal of speculation on the part of scientists, even present realities are sometimes the subject of much scientific confusion. Scientists can't agree on present phenomena like global warming, whether the weather is getting warmer or not and why. Many no longer call it "global warming" but prefer the terms "climate change" or "global climatic disruption". Some meteorologists are now predicting that our planet is headed into a cooling period for the next thirty years or so. This is reminiscent of the 1970's when many scientists were warning of the great disaster facing the earth in a new ice age that was soon to descend on us. Who knows? It may yet come to pass.

The scientific experts can't agree on whether soy is an incredibly healthful product or a substance which is harmful to a person's body. Many of them will gladly pontificate at great lengths on the uselessness of taking vitamins, or going to a chiropractor, or using homeopathic or naturopathic medicines. Others will be equally scientific in promoting the benefits of these therapies to an individual's health. There are a myriad of similar examples. All of these are present realities that they can test and experiment on to their mind's delight and to the health of their bank accounts. Nevertheless, there is a great deal of scientific confusion on these and other tube testable matters.

Why believe as inerrant what some scientists say regarding the age of the earth, where the universe came from, or that a bear or hyena magically evolved into a whale over millions of years, when they are so uncertain about many present realities? Since the past cannot be repeated or experimented on, the evolutionary scenario is based solely on faith in very speculative stories and very limited data, assembled on the back of a materialistic philosophy.

Many scientists are very confident that they are absolutely, or at least ninety-nine percent, right. Self confidence does not establish facts. What if the one percent is right, and the ninety-nine percent wrong? They may be ninety-nine percent confident in their own intelligence and inferences but in the case of evolution they are one hundred percent wrong because their assumptions are wrong. Specifically they reject the reality of GOD, or at least of GOD'S hand in the creation of the universe. The reality of the infinite GOD is the ultimate game-changer when it comes to understanding the creation of the universe and of life, as we will see in the next chapter. GOD says: *"For My thoughts are not Your thoughts", declares the LORD, "neither are My ways your ways. For as high as the heavens are above the earth so are My ways higher than your ways and My thoughts than your thoughts." (Isaiah 55:8,9)*

Having some faith in the reliability of the scientific method for comprehending present realities may be helpful in dealing with many practical aspects of life. However, trusting in historical or futuristic scientific speculations is a category error that will lead to many false beliefs about the past and the future, and thus have a harmful impact on individuals and our world, as we will show in chapter seven.

THE SCIENTIFIC RECORD: GEMS OF TRUTH MIXED WITH FOOL'S GOLD

Why were many scientists of the past arrogantly convinced of certain "facts" that have since been shown to be completely wrong? Darwin, for example, thought the cell was just a blob of protoplasm, like a piece of jelly. He couldn't have been further from the truth, not only with regards to the cell, but also to what the incredible irreducible complexity of the cell reveals concerning the impossibility of the story of evolution which he began. He had no comprehension of how intricately detailed all of life is, beginning with the smallest aspect of life, the cell and the billions of molecules that work together to make the cell work. His rejection of GOD and His Word led him to many false assumptions and therefore false conclusions. His view of evolution was just as erroneous as his view of the cell.

Whatever speculations the scientific experts may make today may be proven completely wrong a few years from now as a result of new information. Today's and tomorrow's discoveries may alter a great deal of what was thought to be true yesterday. It is a good thing to be able to change our minds based on new credible evidence, but why then are they so insistent that what they say today is correct, factual, and trustworthy, when they know it could change tomorrow? No one knows what scientists will believe fifty or a hundred years from now about anything because no one knows what new information may be discovered.

It is not wise to make definitive conclusions based on very limited knowledge. The total knowledge of humanity today is likely only a fraction of the total knowledge regarding our world and universe. The truth is we don't know how small a fraction of the total amount of knowledge it is that we know because we don't have any idea as to how much we do not yet know. How many of our present scientific assumptions will turn out to be a lot of wind with no substance? The history of science demonstrates how little scientists from the past really understood of material realities, even as they gained little bits of knowledge. There are many gems of truth in the history of science but also much alchemy producing fool's gold. Hold on to your scientific "facts" and theories loosely.

WHAT TORNADOES, EXPLOSIONS, WATER AND WIND CAN'T EXPLAIN

Scientific materialism alone is never sufficient for explaining the origin of anything that is intelligently created. Whether it is the computer I am working on, or the words that I am writing, scientific materialism cannot explain the origin of these things. The scientific method may be able to explain the basic laws and processes that make a computer, light, car, or airplane function, but it can never explain the origin of these things from material alone. These amazing inventions did not materialize or evolve out of thin air, all by themselves through the laws of nature and millions of years of tornadoes and volcanic explosions. Ultimately, however, that is exactly what evolutionists believe, because they believe that the intelligence that created all of these products is itself the product of mindless chance over millions of years.

The history and origin of everything that humans have developed is the intelligent use of information to manipulate raw materials to create highly complex and specified products. Material alone is not a sufficient explanation for anything that is created, whether human or divine. Intelligent agency is necessary for all created realities, from the atom to the cosmos, and that is something scientific materialism can never accept. No one can explain a computer without the role of intelligent design, and yet scientific materialists try to persuade people to believe that single cells, more complex than any computer system, came into existence without intelligent design.

A number of creation scientists have used the illustration of the sculptures of the four American presidents on Mount Rushmore. We know that they are the result of intelligent human design and not just some natural forms that have been produced through the process of erosion. Imagine taking little children to see the faces, and, instead of explaining to them that people carved these images, you told them with great awe and wonder that these faces were the result of wind and water erosion that happens every day around us? You might be able to convince the gullible ones, but I suspect most of them would realize, even from their limited experience, that intelligent people probably did it.

We know that rocks are being changed or evolving through erosion into many different shapes, so why not, by sheer chance over millions or billions of years, the shape of human faces that just happen to resemble real people. Apart from historical knowledge, all we can conclude is that these rock formations are presently affected by erosion. If someone brings to the table an *a priori* assumption that there was no intelligent involvement in the sculpturing of these faces, then whatever natural explanation he or she came up with regarding their origin would be false. It would be a fascinating story involving a lot of imagination glorifying the incredible magical power of time and luck working through the molecules of water and wind, but it would have no resemblance to the truth of the creation of these images.

So why do people believe that the faces are carved out? Not because of the scientific analysis of the rocks, but because of the record of history combined with our knowledge and experience that complex sculptures like that do not happen without intelligent design. No matter how many people may try to convince us that they were just the product of evolutionary weathering over time, we know logically that that kind of design and work is impossible without intelligence. The same principle applies to the incredible complexity of all life forms, and indeed, to the universe itself.

The scientific method may be helpful when confined to the study and search for understanding of the present functions and realities of our material world, but it goes astray, far beyond its area of expertise, when it begins to speculate and develop stories regarding the origin of our cosmos and of life from purely materialistic assumptions. Secular materialists need a creation myth to justify their position and are more than happy to prostitute science to help them explain everything without GOD. They come with an a priori bias against the presence and creative activity of GOD. However, wherever unbiased intelligence meets intelligent complex design it will recognize the reality of an intelligent designer. When we come face to face with the incredibly complex reality of our universe and of life, unbiased intelligence will recognize the reality of an infinitely greater intelligence than ours at work.

IS EVOLUTION A MASSIVE WORLDWIDE CONSPIRACY?

Are all of the scientists who believe that evolution is a fact part of a massive worldwide conspiracy to deceive everyone? Would so many qualified scientists around the world believe and teach evolution if it was not scientifically true and accurate? Of course we could ask the same question in many forms: Why would so many millions of people have died in wars for Nazism or Communism if they weren't true? Why did so many of the German intelligentsia, including many scientists, follow Hitler in his madness? Why would Islamic extremists blow themselves up if Allah wasn't true? Why do so many people in India worship cows and other creatures and millions of sculpted idols, if they aren't true? Contrary to their own glorified understanding of themselves, scientists are people, no better, no worse, no wiser than other people, and as easily deceived as anyone else.

No doubt some of those who worship science and scientists will be offended and claim that this is anti-science. This is not anti-science. It is being realistic about the nature of scientists. Peer pressure, following the crowd, the herd instinct, safety in numbers, the desires of the flesh, jealousy, pride, and the love of money and power are only a few of the basic fallen instincts that motivate people, including scientists. To err is human and, as humans, scientists make mistakes, many of them as a result of selfish desires and philosophical biases.

Often those who reject the evolutionary story are accused of believing that all the scientists who affirm evolution are involved in some kind of gigantic secret plot to deceive the whole world. There is no need for a gigantic secret plot by scientists because everyone who chooses to affirm that the sum total of reality is a material universe in which only materialism operates has to believe in some version of evolution.

There is no doubt that, in recent years, many materialists have been very attracted to science, and particularly the evolutionary understanding of life, because to them it is foundational for their whole view of life. If you remove GOD from the equation of life, evolution is the only option that is left. Of course, there are some who start out believing in a materialistic universe and end up seeing the light and turning to GOD, and there are others who begin by believing in GOD and then veer to materialism. Anyone who leaves GOD and His Word out of the final equation, which most scientists do, will be wrong in the end. They may be right about some minor aspects of life, but they will have missed the big picture.

Even though everything from a basic cell to the human mind seems to be incredibly designed beyond anything humanity could create, according to scientific materialism it cannot be designed because that would require a non-materialistic explanation. Dr. Phillip Johnson, in his book *Darwin on Trial,* wrote:

> The National Academy of Sciences told the Supreme Court that the most basic characteristic of science is "reliance upon naturalistic explanations," as opposed to 'supernatural means inaccessible to human understanding'. In the latter, unacceptable category contemporary scientists place not only God, but also any non-material vital force that supposedly drives evolution in the direction of greater complexity, consciousness, or whatever. If science is to have any explanation for biological complexity at all it has to make do with what is left when the unacceptable has been excluded. Natural selection is the

best of the remaining alternatives, probably the only alternative.

In this situation some may decide that Darwinism simply must be true, and for such persons the purpose of any further investigation will be merely to explain how natural selection works and to solve the mysteries created by apparent anomalies. For them there is no need to test the theory itself, for there is no respectable alternative to test it against.[67]

Dr. Eugenie Scott, an atheistic evolutionist, the executive director of the National Center for Science Education, a signer of the Humanist Manifesto #3, and author of *Evolution vs Creationism,* supports this exclusion of God or anything supernatural from science. She writes in the first two paragraphs of her first chapter:

We live in a universe made up of matter and energy, a material universe. To understand and explain this material universe is the goal of science, which is a methodology as well as a body of knowledge obtained through that methodology... What distinguishes science from other ways of knowing is its reliance upon the natural world itself as the arbiter of truth.[68]

To many secularists science is no longer a search for truth but a search for a materialistic explanation. If you remove GOD from the picture and believe in scientific materialism alone, then every piece of data has to fit into a materialistic mold, or in the case of life, into the biological evolutionary mold. There is no other materialistic explanation of life's origins. Consequently, everything proves evolution because it has to. The mathematical reality that evolution is impossible is irrelevant because life exists and therefore evolution happened. In this view evolution is a fact because it is the only materialistic explanation for life, and only materialistic explanations are acceptable. There are many questions for which present day scientific materialism has no answers, but we are assured by those who hold to this position that materialistic answers to these questions will be found. There has to be a materialistic answer because they believe that is all there is.

For those who believe in this understanding of science, every individual scientific discipline including astronomy, biology, geology, chemistry, physics, and palaeontology has to conform to this materialistic and evolutionary mold, which includes billions of years. Every new piece of data discovered has to be forced into the evolutionary mold because it is the only acceptable mold that scientific materialism allows. It isn't surprising, therefore, that any scientist who believes in the doctrines of materialism, regardless of the field they are in, would also believe that evolution is a fact. Thus, there is no need for any conspiracy theory concerning evolution since the evolutionary mold regarding origins is the only one available to those who hold to scientific materialism.

Even if GOD'S signature was encrypted in the genetic codes they would have to explain it with a materialistic cause like a fluke mutation. Indeed, as already stated, the genetic codes are the signature of GOD. The DNA codes reveal a technology, epistemology, and creativity beyond the capabilities of humanity. Any code is the result of intelligence; the genetic code is the result of unimaginable Divine intelligence.

Many scientists certainly don't like the idea of a GOD who may be smarter than they are and whose science is light years higher than theirs. The reality of GOD as our Creator also confronts us with the idea of ultimate accountability to GOD, which is completely anathema for a generation that rejects accountability to any higher authority than themselves.

WHEN HUBRIS LEADS TO BLINDNESS

Many evolutionists argue philosophically that there are numerous poorly designed features and creatures which an intelligent designer would never have created. They cite the following examples: the apparent backward design in the wiring of the human eye, rabbits that eat their own excrement, and a species of woodpecker that has a tongue that winds around in its head for awhile before ending in its mouth. These are not examples of poor design but of pure hubris and ignorance on the part of these evolutionists.

These philosopher scientists cannot even create a single cell or change a house fly into a horse fly, yet some of them are so arrogant that they think they can lecture GOD on how He could have made a better design for our eyes or for the woodpecker's tongue. It would be interesting to hear their tongue twisting explanation as to how the tongue of this woodpecker slowly evolved and what evolutionary advantage their evolving tongue would have had each step of the way, before it got to the mouth. If we are playing the philosopher speculator game, maybe GOD decided to design this particular woodpecker's tongue in such a way to prove that evolution could not have done it. It would take an incredibly creative designer to work out all of the genetic programming and a myriad of other details in creating a woodpecker with that kind of a tongue.

Regarding the electrical wiring of our eyes, these materialistically minded scientists simply don't understand the entire complexity of sight involving the eyes, the brain, and all the connections in between, plus the integrated genetic codes involved in the creation, maintenance and repair of our visual capability. No one but GOD does. They have, as my Dad use to say, "Windt in Kopf"- "wind in the head". They would be well advised to quit blowing hot air about things they do not fully understand. An ounce of humility taken seven times a day would be a great benefit to everyone involved in science and every other human endeavor.

Using the intelligence GOD gave us, we have created incredible technologies and can do amazing surgeries today to help deteriorating or malfunctioning eyes to see better. This is only possible because there already is an incredibly complex system of sight that can be repaired. We could never create a more incredible system of sight than that which has been given to us. To be sure, as a consequence of sin our biological systems no longer work perfectly. Nevertheless, their intelligent design is visible for all but the blind to see.

THE MATERIALISTIC TUNNEL VISION OF SCIENCE

The scientific approach is only applicable to the narrow physical sphere of life. In the purely materialistic world of many evolutionists, life is simply chemical reactions, meaningless molecules mixing and meandering to nowhere. In their world there is no reality to love, joy, peace, and hope. Within the materialistic framework these are nothing but false perceptions created by the chemical and electrical interaction of atoms.

This tunnel vision of scientific materialism results in a denial of everything that is not materially based. If it cannot be put in a test tube, is not made out of chemicals, or is not measurable, it is not part of reality as defined by materialists. Consequently, they believe that GOD, soul, mind, love, joy, and peace are all mere illusions of material. Although doctrinaire secularists proclaim that life is nothing more than evolved material, most people know both by observation and personal experience that life also involves the realities of mind and spirit.

Something to meditate on: can material atoms have illusions? Do molecules experience joy, love, hate, and peace? Do chemicals have feelings? Do electrical impulses make any sense out of the world or think about GOD? Do atoms concern themselves with the ethics or morality of life and death? Do the stars care about you or your family or about the state of the nation? When was the last time you apologized to a tire for kicking it? Why is kicking a dog, or a child, or your neighbour, any different than kicking a tire?

The material reality of atoms cannot produce self-awareness. There is no "self" in a material reality of life, only different arrangements of chemicals. If all that existed was material atoms, then any creature with some degree of self-awareness could never exist. We could never be more than just meaningless molecules, just atoms in a particular form for a small period of time before being rearranged.

The human mind and personality are realities, despite the strictly scientific materialistic tunnel vision perspective. Science, by its very nature, is limited to the physical realities of material and can never know the spiritual reality of GOD or the reality of the nature of man, which involves the spirit and mind as well as the body. We could not know about atoms apart from the reality of the mind. Without the gift of the mind and spirit there would be no science for there would be nothing trying to make any sense of the world. The writing of books, development of experiments, inventions of new technology would all be impossible in a strictly materialist universe because thoughts couldn't exist without minds to think them. In the real world, thoughts, emotions and GOD exist. Pure materialism ignores the experience and observations of billions of people to the realities of life above and beyond material. Those who limit their understanding of reality to the material world are like those who can't see beyond the tip of their own nose.

The scientific focus on the physical realities of life is relevant to our existence in this world and is a gift of GOD. Nevertheless, it is a narrow view of the totality of life. Certainly life has a physical or material aspect to it which is helpful for us to understand. As long as science confines itself, with proper humility, to the recognition of its limited sphere of understanding of present physical realities, it can be a good servant to a humanity that is concerned with the golden rule. However, the materialist view is extremely narrow and destructive when it is seen as being the only truth concerning life.

We live not only in a material universe made up of chemicals and atoms and energy but in a world of thought and spirit: GOD'S, angels' (good and evil) and ours. Since there is an interconnection between body, mind and soul, between the physical universe and the spiritual world of GOD, it is impossible to completely understand the physical world without grasping the basic truths of the spiritual world that GOD reveals through His SON and SPIRIT.

SCIENCE: HELPFUL, NEUTRAL, DESTRUCTIVE, OR ALL OF THE ABOVE?

I am in no way suggesting that science, the ability to experiment and learn about present realities, isn't helpful for human life. Science has brought many wonderful benefits that enable us to see, experience and appreciate the countless blessings of our material world. Science is a great blessing but, and it is a big "but", it can also be extremely harmful and destructive. Nuclear technology can be used to heat homes or to destroy cities and nations. Even the scientific genius Albert Einstein came to regret his own involvement in the development of the atom bomb because of its enormous destructive use.[69] Medical technology can bring healing to the sick or eliminate unwanted people. We can feed the world through crop science or poison or destroy it with chemical and biological weapons. The internet is a useful tool for disseminating much true and useful information, but it is also being used to pollute people's minds with lies and deception that bring moral and social depravity and corruption into people's lives. Many lives are lengthened and improved through scientific advances. The lives of others are diminished or cut short.

Scientific understanding is neutral and can be used for good or evil. According to CHRIST: *"If those days had not been cut short, no one would survive, but for the sake of the elect those days will be shortened." (Matthew 24:22)* We are living in those days in which science has provided humanity with the technological tools that have given the human race the capability of self-destruction. Many secularists are expecting it. The Doomsday clock is never far from twelve midnight.

Scientists are not GOD and science is no savior. Praise GOD He IS! It is time to turn our thoughts and our hearts to GOD, our Creator and Savior!

PART B

THE ONE AND ONLY GOD

Chapter 4

GOD IS!

"I Am Who I Am!" Exodus 3:14
GOD "is the same yesterday, today, and forever"
Hebrews 13:8

What is Infinite?

What is infinite?
  What is eternal?
    That which is infinite from which the finite has come;
      that which is eternal from which the temporary has sprung:
        the great I AM,
          the ALPHA and OMEGA, the BEGINNING and the END,
            the LORD GOD ALMIGHTY, FATHER and SON and HOLY SPIRIT:
              the GOD of Abraham and Sarah, Isaac and Jacob,
                of David, Samuel and the prophets,
                  of Mary, Martha and Lazarus,
                    of Peter, John and Paul,
                      of faithful mothers and fathers,
                      of evangelists, pastors, and priests,
                        of those who have remained single for the Kingdom,
                        of doctors, nurses, teachers, carpenters, and plumbers,
                      of all who have gone before us,
                    who have lived, served, and died by His Word and promises,
                    who have given their lives as a testimony to the ever Living I AM.
                Glory to GOD in the highest
              and on earth peace to all people on whom His grace rests!
          He Who Was and Is and Is to Come says:
        "Yes, I am coming soon. Amen! Come LORD JESUS.
The grace of our LORD JESUS, be with GOD'S people. Amen!"
(Revelation 22:20-21)

KNOWING GOD

The googolplex of complexity of our world and universe deals a powerful knockout blow to the evolutionary scenario, but it is the infinite nature of GOD that is the death knell to this secular myth, as it is to all myths. GOD'S eternal nature is the nail in the coffin of evolution. His infinite, omnipotent, omniscient, holy, and loving nature is the truth that sets us free from the evolutionary mythology.

For those who acknowledge the presence and work of GOD in the created universe, the crucial question becomes, "What is GOD like?" Has our Creator chosen to be a mysterious entity that no one can ever really know? Certainly everyone who sincerely believes in GOD ought to desire to know GOD and to do His will. If you believe that there is a GOD who created you and everything that exists, then your chief goal in life ought to be to come to know the real one and only GOD and to understand why He created you and our universe. As the song *"The Greatest Thing"* by Mark Pendergrass says:

> *"The greatest thing in all of my life is knowing You.*
> *The greatest thing in all of my life is knowing You.*
> *I want to know You more, I want to know You more,*
> *The greatest thing in all of my life is knowing You.*
>   *(second verse: loving You, third verse: serving You)*

There is no greater joy than that which comes from knowing, loving and serving the living GOD. More and more people are coming to experience this joy in their lives every day. Some may ask: How can we know what GOD is really like and what His will is for human life? How do we know if someone who claims to be bringing a message from GOD or about GOD is testifying to the one true living and eternal GOD? The only protection we can have from those who bring false testimonies concerning GOD (there are more false testimonies than true ones) is to truly know GOD for ourselves. But how can anyone know GOD? Is it possible to really know our Creator?

Many have been deceived into believing that it is impossible to really know GOD, and therefore they willingly live their lives in ignorance of their Creator. Anyone who claims to actually know GOD is assumed to be a lunatic, on the same level as someone who claims to have come from the planet Zoon in the outer reaches of some distant galaxy. Of course it would be impossible to know GOD if He didn't want us to know Him. But GOD does want us to know Him. From the beginning to the end of His Word, the Bible, GOD encourages people to sincerely seek Him, and He promises that if they do seek Him they will come to know Him.

It is written:

"Seek the LORD while He may be found; call on Him while He is near. Let the wicked forsake His way and the evil man his thoughts. Let him turn to the LORD, and He will have mercy on him, and to our GOD and He will freely pardon." (Isaiah 55:6,7)

"You will seek Me and find Me when you seek Me with all of your heart. I will be found by you." (Jeremiah 29:13)

"Without faith it is impossible to please GOD, because anyone who comes to Him must believe that He exists and that He rewards those who earnestly seek Him." (Hebrews 11:6)

The reward of seeking GOD is coming to know Him and His love and plan for us. Knowing GOD is a two-way street.

GOD seeks after us while inviting and strongly encouraging us to deliberately seek Him. It is written:

> *"This is eternal life that they may know You, the only true GOD, and JESUS CHRIST, Whom You have sent" (John 17:3)*

> *"We know also that the SON of GOD has come and has given us understanding, so that we may know Him who is true. And we are in Him who is true— even in His SON JESUS CHRIST. He is the true GOD and eternal life." (1 John 5:20)*

When people come to know GOD they soon realize that GOD wants us to know Him and has revealed Himself to us through His SON, His SPIRIT and His Word. The whole life of CHRIST and the whole of the Bible are given to us so that we can come to know the one true and living GOD and live our lives in His presence, now and for all eternity.

GOD is the central reason for rejecting the evolutionary mythology. GOD is GOD! He is who He IS! He is who He has revealed Himself to be. It is not for us to make GOD into our image or according to our imagination, or to limit GOD by a narrow materialistic perspective. The argument in the Christian Church today over evolution is not just about a "minor" difference in the interpretation of the Hebrew word *"yom"*, translated as "day" in Genesis 1. The controversy is over the very nature and character of GOD.

GOD IS!

In order to understand why the nature of GOD is contrary to the process of evolution we need to have some clear comprehension of the nature or character of GOD. The core reason evolution has crept into the Christian church under the guise of science is largely because many people lack a clear and consistent CHRIST-centered and Biblical understanding of who GOD is. Since most people do not know the nature of GOD, they are easily led into a false conception of GOD that would fit in with the contemporary idea of evolution. The temptation, contrary to the First Commandment, is to make GOD into an image acceptable to us and to the world around us, a god who makes us comfortable. What we need, of course, is to know the true living GOD, our Creator.

The Christian understanding of the nature and character of GOD has always been based on GOD'S self-revelation through JESUS and through the HOLY SPIRIT in His Written Word. So what has GOD revealed about Himself through CHRIST, the SPIRIT and the Word?

GOD IS ETERNAL!

1. *"Abraham planted a tamarisk tree in Beersheba, and there he called upon the name of the LORD, the Eternal GOD." (Genesis 21:33)*

   "The eternal GOD is your refuge, and underneath are the everlasting arms." (Deuteronomy 33:27)

If GOD was mortal, rather than eternal, we would not be able to rely upon His help or rest in His arms.

2. *"You have made known to me the path of life; You will fill me with joy in Your presence, with eternal pleasures at Your right hand." (Psalm 16:11)*

> *"Surely You have granted him eternal blessings and made him glad with the joy of Your presence." (Psalm 21:6)*

GOD cannot give what He does not have. If GOD did not have an eternal nature, He could not give eternal life or eternal blessings as He promises repeatedly in His Word. Everyone who sincerely believes that GOD can give them eternal life has to believe that GOD is eternal.

> **3.** *"The fear of the LORD is the beginning of wisdom; all who follow His precepts have good understanding. To Him belongs eternal praise." (Psalm 111:10)*

GOD could not receive eternal praise if He was not eternal.

> **4.** *"Your Word, O LORD, is eternal; it stands firm in the heavens." (Psalm 119:89)*
>
> *"All Your Words are true; all Your righteous laws are eternal." (Psalm 119:160)*

GOD'S Word and law could not be eternal if GOD was not eternal.

> **5.** *"Trust in the LORD forever, for the LORD, the LORD, is the Rock eternal." (Isaiah 26:4)*
>
> *"He does not change like shifting shadows." (James 1:17)*

GOD is worthy of trust only because He is eternal and never changes.

> 6. *"Do not work for food that spoils, but for food that endures to eternal life, which the Son of Man will give you. On Him GOD the Father has placed His seal of approval." (John 6:27)*
>
> *"For my Father's will is that everyone who looks to the Son and believes in Him shall have eternal life, and I will raise him up at the last day." (John 6:40)*

Anyone who reads the Gospels on the life of JESUS knows that He was always talking to people about eternal life and the eternal Kingdom of GOD. If GOD was not eternal, JESUS could not have promised eternal life to anyone.

> 7. *"Now to the King eternal, immortal, invisible, the only GOD, be honor and glory forever and ever. Amen." (1 Timothy 1:17)*
>
> *"How much more, then, will the blood of CHRIST, who through the eternal Spirit offered Himself unblemished to GOD, cleanse our consciences from acts that lead to death, so that we may serve the living GOD!" (Hebrews 9:14)*

Contrary to Mormon doctrine, the clear self-revelation of GOD is that He is eternal. GOD wouldn't be GOD if He wasn't eternally GOD.

GOD IS ALMIGHTY!

> 1. *"When Abram was ninety-nine years old, the LORD appeared to him and said, "I am GOD Almighty; walk before Me and be blameless."*

> *(Genesis 17:1)*
>
> *"May GOD Almighty bless you and make you fruitful and increase your numbers until you become a community of peoples." (Genesis 28:3)*

If we believe in the same GOD Abraham believed in then we know that GOD is almighty.

> **2.** *"Will the one who contends with the Almighty correct Him? Let him who accuses God answer Him!" (Job 40:2)*

GOD is called the Almighty over thirty times in the book of Job, probably the oldest book in the Bible.

> **3.** *"Of the increase of His government and peace there will be no end. He will reign on David's throne and over His kingdom, establishing and upholding it with justice and righteousness from that time on and forever. The zeal of the LORD Almighty will accomplish this." (Isaiah 9:7)*

The prophet Isaiah calls GOD the Almighty over sixty times.

> **4.** *"This is what the LORD Almighty, the GOD of Israel, says: 'I will break the yoke of the king of Babylon." (Jeremiah 28:2)*

The prophet Jeremiah calls GOD the Almighty over eighty times, and the remaining prophets in the Old Testament call GOD the Almighty over one hundred times. The HOLY SPIRIT repeatedly calls GOD the Almighty, in sharp contrast to all of the powerless idols of the world.

> 5. "For nothing is impossible with God." Angel Gabriel (Luke 1:37)

The only way that nothing would be impossible for GOD to do is if He was and is and always will be almighty.

> 6. "Jesus looked at them and said, 'With man this is impossible, but with GOD all things are possible.'" (Matthew 19:26)

JESUS is talking here about the gift of salvation. GOD could only guarantee eternal salvation to people if He is almighty. In His life, ministry, and resurrection JESUS demonstrated the infinite power of GOD.

> 7. "I am the Alpha and the Omega," says the Lord God, "who is, and who was, and who is to come, the Almighty." (Revelation 1:8)
>
> "Each of the four living creatures had six wings and was covered with eyes all around, even under his wings. Day and night they never stop saying: "Holy, holy, holy is the LORD GOD Almighty, Who Was, and Is, and Is to come." (Revelation 4:8)
>
> "Then I heard what sounded like a great multitude, like the roar of rushing waters and like loud peals of thunder, shouting: "Hallelujah! For our LORD GOD Almighty reigns." (Revelation 19:6)

The Almighty reigns; He always has, and He always will. It is good and right for us to praise Him because of who He is, what He has done, and what He promises to do. Hallelujah! Praise the LORD! He is worthy of all of our praise.

GOD IS INFINITE IN INTELLIGENCE AND WISDOM!

> *1. "To God belong wisdom and power; counsel and understanding are His." (Job 12:13)*

GOD is the only source of true wisdom and understanding.

> *2. "How many are your works, O LORD! In wisdom you made them all; the earth is full of your creatures." (Psalm 104:24)*

All of the creatures of the earth are a clear and living testimony to the intelligence and creativity of GOD.

> *3. "But GOD made the earth by His power; He founded the world by His wisdom and stretched out the heavens by His understanding." (Jeremiah 10:12)*

GOD'S wisdom here is in reference to His creation of the universe and world. If the universe and all life forms were not originally created perfectly then GOD is not infinitely wise, in which case He would not be the GOD testified to in the Bible. GOD'S wisdom has to be infinite, because He is infinite.

> *4. "All this also comes from the LORD Almighty, wonderful in counsel and magnificent in wisdom." (Isaiah 28:29)*

GOD'S knowledge and wisdom revealed in all that He does and says is absolutely amazing to those who begin to grasp even a small fraction of it.

> 5. *"For My thoughts are not your thoughts, neither are your ways My ways," declares the LORD. "As the heavens are higher than the earth, so are My ways higher than your ways and My thoughts than your thoughts." (Isaiah 55:8,9)*

GOD'S intelligence, understanding, and ways of accomplishing His will are light years beyond us.

> 6. *"Oh, the depth of the riches of the wisdom and knowledge of GOD! How unsearchable His judgments, and His paths beyond tracing out!" (Romans 11:33)*

GOD'S thoughts are not only higher than ours, they are also much deeper. His wisdom and knowledge are beyond our full comprehension.

> 7. *"CHRIST, in whom are hidden all the treasures of wisdom and knowledge." Colossians 2:3*

All wisdom and knowledge are found in CHRIST, Who is GOD!

GOD IS HOLY AND PERFECT!

> 1. *"I am the LORD your GOD; consecrate yourselves and be holy, because I am holy." (Leviticus 11:44)*

This is part of GOD'S self-description of who He is. He is completely holy.

> 2. *"He is the Rock, His works are perfect, and all His ways are just. A faithful GOD who does no wrong, upright and just is He." (Deuteronomy 32:4)*
>
> *"As for God, His way is perfect; the Word of the LORD is flawless." (2 Samuel 22:31)*
>
> *"Sing to the LORD, you saints of His; praise His holy name." (Psalm 30:4)*

GOD'S perfect works include our universe and world prior to humanity's fall into darkness. His works are perfect because He is perfect. If He didn't do everything perfectly, He wouldn't be perfect. GOD is worthy of praise because He is holy and perfect.

> 3. *"Do not cast me from your presence or take your Holy Spirit from me." (Psalm 51:11)*

GOD'S SPIRIT is called the HOLY SPIRIT over one hundred times in the Bible.

> 4. *"You are good, and what You do is good" (Psalm 119:68)*
>
> *"Why do you ask me about what is good?" Jesus replied. "There is only One who is good." (Matthew 19:17)*

This isn't saying that GOD is just a little good or that He does a little good, but that He is good- perfectly good and what He does is pure goodness. JESUS says that only GOD is good- pure unalloyed goodness. As the Apostle John put it in his letter: *"GOD is light; in Him there is no darkness at all" (1 John 1:5)*

5. *"The fear of the LORD is the beginning of wisdom, and knowledge of the Holy One is understanding." (Proverbs 9:10)*

This is very applicable to understanding the creation-evolution controversy. Without knowing the holiness, wisdom and power of GOD people cannot have a proper understanding of the truth concerning our origins.

6. *"The angel answered, "The HOLY SPIRIT will come upon you, and the power of the Most High will overshadow you. So the Holy One to be born will be called the SON of GOD." (Luke 1:35)*

   "We believe and know that you are the Holy One of GOD." (John 6:69)

The Holy One referred to here is JESUS CHRIST, GOD'S SON.

7. *"Holy, holy, holy is the LORD Almighty; the whole earth is full of His glory." (Isaiah 6:3)*

   "Each of the four living creatures had six wings and was covered with eyes all around, even under his wings. Day and night they never stop saying: "Holy, Holy, Holy is the LORD GOD Almighty, Who Was, and Is, and Is to Come." (Revelation 4:8)

The angels declare the holiness and glory of GOD revealed in His eternal nature.

In addition to the hundreds of times in Scripture that GOD is called holy, there are hundreds of other times that everything associated with GOD is called holy: Holy Scriptures, Holy Name of GOD, Holy City, Holy Temple, Holy people, Holy angels, Holy Sabbath, Holy Kingdom, and Holy mountain.

ONE GOD, MANY IDOLS

"I am the LORD your GOD...You shall have no other gods before Me. You shall not make for yourself an idol in the form of anything in heaven above or on the earth beneath or in the waters below. You shall not bow down to them or worship them." (Exodus 20:2-4)

GOD is eternal, almighty, infinitely wise, and perfectly holy. In chapters 5 and 6 we will examine the truth that GOD is perfect love and perfect truth: He never lies and His Word is truth. If people do not believe that GOD'S essential nature is all of these things, then they believe in a different GOD than the GOD of Abraham, Jacob, David, the prophets, and the apostles. If people consider themselves to be Christian believers, but don't believe that GOD is perfect in all of these ways, then *"Houston, we have a problem"*. This is on par with the Barna poll, taken in 2009, that asked people who considered themselves to be Christians if they believed that JESUS was sinless (perfect-holy). Approximately one-third of the people who claimed to be Christians did not believe that JESUS was sinless.[70] They believed that He was a sinner just like the rest of us, which is a clear indication that people do not know CHRIST and do not have even a basic understanding of the Christian faith. Likewise, anyone who doesn't believe that GOD is perfect doesn't know who GOD is.

The human tendency is to make GOD in our own fallen image. For many people, GOD is just a little more powerful, a little smarter, and maybe even a little better or holier than we are. However, GOD isn't just a little stronger, smarter or better than us. He is the infinite GOD, who is omnipotent, omniscient, and perfectly holy in every way. All of these attributes of GOD are perfectly integrated with each other. GOD is who He is.

CAN SOMEONE BELIEVE IN CHRIST AND EVOLUTION?

Within the context of human history and human imagination there are a myriad of different philosophical or religious views of GOD. By some He is viewed as an absentee landlord or a curious bystander watching and wondering how it will all turn out. To others He is not good or wise or infinite in power. Tragically, these views of GOD are often being held and spread by many false teachers and false prophets in the institutional church. At one time they, or at least their faith ancestors, knew CHRIST and His Word, but they have now rejected the revelation of CHRIST, and are dancing in unison with the world's many caricatures and idols of GOD.

Numerous philosophical views of GOD including deism, Buddhism, Hinduism, animism, and many of the new age spiritualities can and do accommodate the evolutionary story relatively well. The HOLY SPIRIT, CHRIST-centered, Biblically inspired revelation of GOD isn't one of them. It is certainly possible for someone to believe in evolution if they hold to one of the false views of GOD, but it is inconsistent for any one holding to the Biblical, CHRIST-centered revelation of GOD to incorporate the evolutionary story. This explains why those within the Christian church who succumb to the evolutionary mythology have to try to transform GOD into a "dumbed down humanized revised version" that the world finds compatible and inoffensive.

It is important to be clear at this point. It is not impossible for someone to believe in JESUS and at the same time believe in evolution. Similarly, just because someone believes in the creation account in Genesis does not mean that he or she is a Christian. GOD alone knows and will judge all hearts. We are not in this world to be judges but to be witnesses of the Living GOD and His eternal realm. I believe that there are many who sincerely believe in CHRIST who are also convinced that evolution is at least partially true.

In the lives of all believers there are inconsistencies in our understanding of GOD, CHRIST, and faith, and a failure to integrate our faith into some areas of our lives. As it is written: *"Now we see but a poor reflection as in a mirror; then we shall see face to face. Now I know in part; then I shall know fully, even as I am fully known." (1 Corinthians 13:12).* Part of maturing in our Christian faith is to grow in our understanding of GOD and His will and Kingdom that He has revealed through His SON, His HOLY SPIRIT and His Word. There is plenty of room for all of us to grow. No Christian understands it all, or has perfect faith in all of the revelations of GOD'S Word.

As with all aspects of life and faith, every sincere follower of JESUS needs to carefully read, think and test the implications of the Biblical revelation of CHRIST versus the god of the theistic evolutionary story. The Apostle John wrote:

> *"Dear friends, do not believe every spirit, but test the spirits to see whether they are from God, because many false prophets have gone out into the world. This is how you can recognize the Spirit of God: Every spirit that acknowledges that Jesus Christ has come in the flesh is from God, but every spirit that does not acknowledge Jesus is not from God. This is the spirit of the antichrist, which you*

have heard is coming and even now is already in the world. You, dear children, are from God and have overcome them, because the one who is in you is greater than the one who is in the world. They are from the world and therefore speak from the viewpoint of the world, and the world listens to them. We are from God, and whoever knows God listens to us; but whoever is not from God does not listen to us. This is how we recognize the Spirit of truth and the spirit of falsehood. Dear friends, let us love one another, for love comes from God. Everyone who loves has been born of God and knows God." (1 John 4:1-7)

There are many false teachings and teachers or prophets in our world who are spreading many fabrications about GOD. In these verses GOD'S HOLY SPIRIT instructs and even commands us to test these teachings and teachers using three criteria:

1. Do they believe in and bring a clear testimony to JESUS CHRIST as GOD'S SON who came into our world as a human being to save us? If they don't believe and teach this central truth, then their spirit is that of the antichrist who opposes CHRIST and seeks to replace him with false saviors.
2. Do they believe and accept the testimony of the prophets and apostles in the Old and New Testament Scriptures as having come from GOD'S HOLY SPIRIT, and is their teaching in harmony with the message of the prophets and apostles? If not, then don't listen to them; they are false guides.
3. Is their teaching in accordance with the love of GOD revealed through CHRIST? If their teaching and lives are contrary to the reality of GOD'S love, then don't

listen to them; their motive and message are not from GOD.

It is crucial for everyone who wants to know and serve the eternal GOD to ask whether the revelation that GOD gives us of Himself through His SON and His SPIRIT in His Word is consistent with the process of evolution. We all need to carefully listen and meditate on the Biblical revelation of GOD versus the god who would be if he was the instigator and tinkerer of evolution.

It is a great error to believe that just because a person claims to have faith in CHRIST, it doesn't matter what other beliefs that person may hold. There are many who claim faith in CHRIST but who hold racist attitudes, believe in reincarnation, love money or sports or pleasure more than GOD, worship idols, abuse their families, engage in immoral relationships, or cheat their customers. The attitude of every sincere believer and follower of CHRIST is to allow GOD'S Word in all of its grace and truth to transform our lives so that we will more clearly reflect the whole character and truth of GOD to our lost world. In each of our lives we need to follow through on what the Apostle Paul inspired by GOD'S SPIRIT wrote to Timothy: *"Watch your life and doctrine closely. Persevere in them, because if you do, you will save both yourself and your hearers."* (1 Timothy 4:16)

What we believe has tremendous consequences, not only for our own lives, but for our world. Those who claim the Name of CHRIST will either be faithful witnesses to GOD or will bring dishonour to the name of CHRIST through their false beliefs and practices. Mixing the teachings of CHRIST, of GOD, His SPIRIT and His Word, with the various mythologies believed and practiced in the world around us is a common error by many who claim CHRIST as their Savior. That path, however, leads away from faith in GOD and His Word and toward the darkness, suffering and hopelessness so visible in our world today. CHRIST made it clear that He will hold each person accountable, and this includes our beliefs and our actions.

IF GOD WERE MOLOCH, EVOLUTION COULD BE HIS MODUS OPERANDI

If GOD were some weak, ignorant, and petty god- a relative of the pantheon of gods of the ancient Greeks and Romans- then evolution would be a good fit. If GOD resembled the ancient near-eastern idols of Baal or Ashtoreth or Moloch then evolution could have been His *modus operandi*. Praise GOD that He is not weak, ignorant or petty! These idols have come and gone into the ash heap of human history! That's not who GOD is or was or will be. GOD is, as revealed through His SON, His SPIRIT and His Word, the infinite, almighty, all wise, holy, and loving GOD. He always was and always will be! As it is written: **"JESUS CHRIST is the same yesterday, today, and forever!" (Hebrews 13:8)**

If GOD didn't create everything perfectly, He wouldn't be perfect. If He didn't do everything justly, He wouldn't be just. If His motivation for everything wasn't love, He wouldn't be love. GOD is not schizophrenic in His nature or character. He is completely and perfectly integrated in all aspects of His being. His loving, truthful, omnipotent, omniscient, holy nature is all in perfect harmony. GOD is consistent and faithful to Himself, to His Word and to His people.

Evolution is either a completely godless mythology, which it certainly is for many people today, or it presents a completely false view and understanding of GOD. There is no resemblance between the kind of deity that would have begun and/or tinkered with the evolutionary process and the living GOD revealed in His Word. Contrary to the evolutionary apologists in the institutional church, evolution is not some grand, elegant, incredibly marvellous and delightful process. It is a slow, painful, and cruel process from one imperfection and error to another. No one in the scientific community would say that evolution is a perfect process. It is a process which could only have been conceived by a group of confused and/or angry scientists, like Charles Darwin.

WHAT IS WRONG WITH THEISTIC EVOLUTION?

The basis of evolution is the belief that everything very slowly evolved from simple to complex. This is an oxymoron today, for we know that there is no universe, or galactic system, or life form that is simple. All life forms, either encased as fossils or living today, are highly complex creatures with incredibly sophisticated, computerized, and marvellously miniaturized programs. Since it is obvious to everyone that life isn't simple, many people have been misled into thinking that the answer is to combine belief in evolution with the belief in a god who, at least, initiated the whole process. This is called "theistic evolution" and it claims to bridge the gap between science and religion.

Theistic evolution is the belief held by many religious people that there is a mystical hidden god who is a behind-the-scenes creator of our universe, world, and life through the process of cosmic and biological evolution. Theistic evolution seeks to harmonize religious belief in a god and the theory of evolution. It claims to be a peacemaker between evolutionary "science" and people of religion, especially Christianity. Many people in churches find theistic evolution very soothing since it appears to remove the stress of having to deal with the conflict between "science" and Christian faith. The comfort of compromise is an easier path than the challenge of the cross to be faithful to GOD. Theistic evolution never quite manages to deliver the peace of mind it promises, since it avoids the conflict between who GOD is and what the god of evolution would be like. As Jeremiah declared: **"They dress the wound of my people as though it were not serious. 'Peace, peace,' they say, when there is no peace."** (Jeremiah 6:14)

Theistic evolution is an untenable position to hold. It is contrary to evolution since it inserts a divine being into an otherwise meaningless chance process. Dr. Paul Zimmerman writes:

> Before going into this, it is worthwhile to note that the hope for reconciliation between the theory of evolution and Christianity as a supernatural religion is not realized by the compromise of saying that evolution is GOD'S way of creating. If you say that, you then inject into the scientific theory of evolution a supernatural factor…The theory is based on the interaction of matter with matter. It is based on the changes which are produced by chance and which are then developed by natural selection. If one places God's guidance into the process, he violates one of the basic tenets of the theory. Moreover, evolution has no place for a man starting in a good world, a man starting with a knowledge of righteousness and true holiness (Eph. 4:24; Col. 3;10). Man does not fall from a lofty position. Rather he is climbing upward under his own power from a lower position and achieving a higher one.[71]

Theistic evolution is also incompatible with Biblical theism since it is not in harmony with the eternal, almighty, omniscient, holy nature of GOD. Theistic evolutionists believe that there is a god who began the whole creative process of our universe. However, in their evolutionary position many say that after igniting the spark of the universe GOD no longer took a direct part in the development of the cosmos or of life itself, but let the laws of "mother nature" run their course. Having credited GOD with the initial spark, many theistic evolutionists are, like their secular counterparts, content to let chance and time do all of the creating.

Since theistic evolutionists believe that there is a GOD who is the originator of our universe, there is absolutely no reason to believe that GOD would have left the scene after initiating creation. Once you acknowledge that GOD began the cosmos, you cannot then exile Him from His creation by insisting that He did not or would not take a direct hand in the development of His universe. If they believe that GOD was directly involved with the initial spark of creation, it is illogical for them to insist on eliminating GOD as a possible explanation for every part of the creative process. If they acknowledge the reality of GOD, they should also be willing to acknowledge that He is free to do whatever He wants in accordance with His Holy nature, and that He cannot be limited to the processes of the known natural laws of twenty-first century science. Once the presence of GOD in the creative process is affirmed, then materialism alone can never be a complete explanation for anything in our universe.

The secular materialist position is at least consistent at this point. It rejects GOD and believes that everything just happened all by itself. The theistic evolutionary position is completely inconsistent. It claims to believe in GOD but at the same time limits GOD to creating the spark for the big bang and perhaps occasionally throwing in a few mutations here and there. In order to accommodate their secular scientific contemporaries, theistic evolutionists have reduced GOD to nothing more than a mystic force out there someplace who has no personal identity or plan for the universe or for human life. Theistic evolution is an illegitimate attempt to bridge the gap between religion and evolution by reducing GOD into the ghost of a deity who is present to haunt us but who chooses not to do anything. The god of theistic evolution has no resemblance to the GOD who IS as revealed through His SON, HOLY SPIRIT and Holy Word.

THE LIVING GOD IS NOT A LITTLE TINKERER

An imaginary god with limited intelligence, resources, and power is the only kind of god that could conceivably have experimented with evolution. The real GOD, who is eternal and infinite, has no need to experiment for billions of years before creating a basic life form, and then take a few hundred million more years to evolve more complex creatures, each of which is weak and subject to all manners of disease, violence and death. This little tinkerer of a god (Little Tinkerer) still hasn't perfected anything even after the imaginary billions of years.

In the evolutionary tale, life is not any more complex, diverse, or efficient today than in the mythical dinosaur era one hundred million years ago. There is just as much imperfect life in our world today as there was in their imaginary prehistoric world, perhaps even more considering the increasing impact of accumulated mutations. Their Little Tinkerer would be an extremely slow learner who is still repeating the same tragic mistakes and evolving the same kinds of weak creatures it has always evolved. This flawed and limited process could not be the creation of a GOD who is holy and has infinite power and wisdom. GOD is an incredible Creator, not a little tinkerer.

DOES THE END JUSTIFY THE MEANS?

The key to understanding theistic evolutionary thinking, besides the need to compromise and have peace with the "scientific community", is the philosophy of "the end justifies the means". This has become a dominant theme in our contemporary secular world as it has turned away from the GOD who IS. "The end justifies the means" is the immoral logic our world uses to justify any and all self-centered human actions. Theistic evolutionists believe that the means or process of evolution is completely justified because of the "end" it produces. Since you and I are living, and because many people believe in their own goodness, therefore, they believe that evolution is a good process. If it produces such "noble" creatures as you and me and the awesome wonders of nature it has got to be good. Case closed! The case is closed, but the case is a casket containing our corpse and the corpse of evolution. Evolution doesn't end in us; it ends in death. We aren't good, and evolution is a fatal and cruel process as we shall see more clearly in the next chapter.

"The end justifies the means" is not the policy of the Holy GOD. To repeat what was stated earlier, everything GOD does is holy and perfect because He is holy and perfect. With GOD the means and the methods are always perfectly consistent with the goal or purpose that He is accomplishing. The end does not justify the means and the means do not justify the end. Both the means GOD uses and the end He accomplishes are perfect and holy. The evolutionary process is neither perfect nor holy and therefore could not be the *modus operandi* of the Holy I AM.

The Little Tinkerer god(dess) of theistic evolution is not like the living GOD revealed through His Son, HOLY SPIRIT and Word. Instead it would be a lot like us. It is understandable that the god or gods of theistic evolution resemble fumbling corrupt human beings. This is one of humanity's specialities: creating gods or goddesses in our own messed-up image, instead of humbling ourselves before the infinite GOD who Is. The god of human choice is a god we can manipulate for our own purposes, a little tinkerer who operates on our kind of ethics, anything but a Holy perfect GOD. A goddess who follows the twisted trails of "the end justifies the means" would be a lot like the human race that always seeks to justify its actions on the basis of what we believe are our "good" goals. Good, of course, meaning what is good for us, our family, our bank accounts, our race, our nation, or our vision of what the world should be like. A quintessential representation of this is Hitler's Third Reich. "The end justifies the means" is used to justify all kinds of evil in our world today. It is a philosophy and an ethic from hell, not from Heaven.

UNBELIEF IN THE SCRIPTURES LEADS TO IGNORANCE OF GOD

Can anyone in their wildest imaginations really believe that the Holy GOD, who is omnipotent and possesses all knowledge, is the creator of a system which is as imperfect and inefficient as the evolutionary process would be? Human beings can create things but they last for just a little while before they fall apart. A brand new computer, as complex as it is, will last for only a few short years. If we had perfect knowledge and power we would live forever and be able to create things which would be eternal like us. If we understood all the intricate dynamics of quantum mechanics and the unknown mysteries beyond it, we wouldn't need cars or planes to transport us to different places. We would just pop in and out of wherever we wanted to be, just like the angels. If we knew all of the intricacies of the human body, we could cure all malfunctions and diseases. We could find that elusive elixir of the fountain of youth. If we had complete knowledge, we could create our own suns and planetary systems that would never run down. Is it possible that the eternal GOD who created our universe is ignorant of any of these realities? Scientists today are already boasting of what they will be able to do fifty or one hundred years from now. Does anyone really believe that the GOD who created our DNA doesn't understand as much as we do, or is less efficient than we are, or couldn't do as much as humans could do one hundred years from now?

From my experience the overwhelming majority of those who believe in the god of theistic evolution inevitably don't believe in the Holy GOD who is infinite in knowledge and power. On the contrary, they tend to believe that Little Tinkerer is evolving, changing, and growing together with the world and universe it began. Their faith and hope is that Little Tinkerer may be evolving to be as good as they think they are. Such is the god they choose to believe in, a god in their own image.

JESUS said to the very liberal unbelieving theologians of His day: *"You are in error because you do not know the Scriptures or the power of GOD." (Matthew 22:29)* JESUS didn't mean that they hadn't read or heard what the Scriptures testify about GOD, but rather that they had chosen not to believe what the Scriptures said. This is also true of theistic evolutionary theologians today. It isn't that they are ignorant of what GOD has written in the Bible, but simply that they have chosen not to believe what GOD has said and revealed in all of the Scriptures is true. JESUS said to His own followers: *"How foolish you are, and how slow of heart to believe all that the prophets have spoken!" (Luke 24:25)* Many don't know the eternal, omnipotent, omniscient, and Holy GOD revealed by GOD'S SPIRIT because they have chosen not to believe what the HOLY SPIRIT has inspired to be written.

The end result of belief in evolution is that people become ignorant of the eternal GOD, for they no longer believe in GOD'S revelation of Himself through His SON, His SPIRIT and His Word. Many theistic evolutionists have chosen to try to remake GOD in their own fallen image. Their Little Tinkerer god, however, exists only in their finite imaginations. Little Tinkerer is a temporary aberration. GOD Is infinite! GOD is also love as we shall see in the next chapter.

Chapter 5

GOD IS LOVE & LIFE

"God is love!" 1 John 4:8
"I have come that they may have life, and have it to the full."
JESUS, the CHRIST (John 10:9-11)

PART A: The GOD of Love and Life or The Little Tinkerer of Misery and Death?

THE LOVE OF GOD

1. The love of God is greater far
   Than tongue or pen can ever tell;
   It goes beyond the highest star,
   And reaches to the lowest hell;
   The guilty pair, bowed down with care,
   God gave His Son to win;
   His erring child He reconciled,
   And pardoned from his sin.

   > Chorus:
   > Oh, love of God, how rich and pure!
   > How measureless and strong!

> It shall forevermore endure—
> The saints' and angels' song.

2. When years of time shall pass away,
   And earthly thrones and kingdoms fall,
   When men who here refuse to pray,
   On rocks and hills and mountains call,
   God's love so sure, shall still endure,
   All measureless and strong;
   Redeeming grace to Adam's race—
   The saints' and angels' song. (Chorus)

3. Could we with ink the ocean fill,
   And were the skies of parchment made,
   Were every stalk on earth a quill,
   And every man a scribe by trade;
   To write the love of God above
   Would drain the ocean dry;
   Nor could the scroll contain the whole,
   Though stretched from sky to sky (Chorus)

-Frederick M. Lehman, 1917

According to GOD the FATHER, JESUS the SON, the HOLY SPIRIT, and the Word, GOD is eternal, almighty, infinitely intelligent, wise, and holy, and GOD is life and love. All life and genuine love is a gift of GOD through creation and through His SON and HOLY SPIRIT. It is GOD who has created us with the capacity to both receive His love and to respond in love to Him and to our neighbor. GOD is overflowing with life and love; evolution is saturated with suffering, evil and death. Before focusing on how GOD'S love dispels the evolutionary fog, it is important to listen carefully to GOD'S revelation of His love through CHRIST, the HOLY SPIRIT and His Word.

It is written:

1. "In Your unfailing love You will lead the people You have redeemed. In Your strength You will guide them to Your holy dwelling." (Exodus 15:13)

   "And he passed in front of Moses, proclaiming, "The LORD, the LORD, the compassionate and gracious GOD, slow to anger, abounding in love and faithfulness." (Exodus 34:6)

   "Know therefore that the LORD your GOD is GOD; He is the faithful GOD, keeping His covenant of love to a thousand generations of those who love Him and keep His commands." (Deuteronomy 7:9)

Frequently GOD'S love is described with the word "unfailing". If GOD'S love was not constant, we could never rely on Him. GOD'S love can be a daily experience in our lives as we live in this love relationship with GOD. GOD has a covenant of love—a loving desire and a voluntary, enduring commitment to love all those who will trust in Him.

2. *"Have mercy on me, O GOD, according to Your unfailing love; according to Your great compassion blot out my transgressions." (Psalm 51:1)*

   "You are forgiving and good, O LORD, abounding in love to all who call to You." (Psalm 86:5)

It is only on the basis of GOD'S love and compassion that we can be forgiven.

3. *"Praise the LORD, all you nations; extol Him all you peoples. For great is His love toward us, and the faithfulness of the LORD endures forever. Praise the LORD." (Psalm 117)*

"Give thanks to the LORD, for He is good. His love endures forever." (Psalm 136:1)

There are only two verses in Psalm 117, the shortest chapter of the Bible. They are a call to all people to praise GOD because of His love. This chapter is found at the very center of the Bible. All twenty-six verses of Psalm 136 contain the words: *"His love endures forever."*

4. *"The LORD is gracious and compassionate; slow to anger and rich in love. The LORD is good to all, He has compassion on all He has made...The LORD is faithful to all His promises and loving toward all He has made...You open Your hand and satisfy the desire of every living thing. The LORD is righteous in all of His ways and loving toward all He has made." (Psalm 145:8,9,13,17)*

Three times in this psalm GOD highlights His love and compassion for every living thing that He has made.

5. *"For GOD so loved the world that He gave His one and only SON, that whoever believes in Him shall not perish but have eternal life." (John 3:16)*

This is the most well known verse in the Bible and it brings a clear message of GOD'S love for people through His SON JESUS.

6. *"I in them and You in Me. May they be brought to complete unity to let the world know that You sent Me and have loved them even as You have loved Me." (John 17:23)*

"And I pray that you, being rooted and established in love, may have power, together with all the saints, to grasp how wide and long and high and deep is the love of CHRIST, and to know this love that surpasses knowledge—that you may be filled to the measure of all the fullness of GOD." (Ephesians 3:17-19)

In JESUS' prayer the night before He was crucified, and in the Apostle Paul's prayer for believers, it is clear that it is GOD'S will for each of us to experience His great love.

7. *"This is how we know what love is: JESUS CHRIST laid down His life for us. And we ought to lay down our lives for our brothers." (1 John 3:16)*

"Dear friends, let us love another, for love comes from GOD... Whoever does not love does not know GOD, because GOD is love...So we know and rely on the love GOD has for us. GOD is love. We love because he first loved us!" (1 John 4:7,8,16,19)

"To Him who loves us and has freed us from our sins by his blood, and has made us to be a kingdom and priests to serve his God and Father—to him be glory and power for ever and ever! Amen." (Revelation 1:6)

Through CHRIST'S death on the cross we can catch a glimpse of how great GOD'S love is for us and how we ought to love others. If GOD isn't love, then there is no ultimate reason for anyone to love GOD or anyone else. GOD is forever worthy of our love and praise and worship because of His love for us through CHRIST. This is our GOD! Love is at the very core of His being.

LITTLE TINKERER: A SUPREME MAD SCIENTIST AND GORE ADDICT

The love of GOD that gives life to everything is being sacrificed today on the altars of theistic evolution. GOD is love and life. Evolution is the exact opposite, it is saturated with suffering, evil and death. Contrary to the evolutionary slogan of *"the survival of the fittest"*, nothing survives for very long in the evolutionary process. Suffering, disease, violence, and death are all part of the very fabric of evolution. Evolution is *"red in tooth and claw"* as Dr. Richard Dawkins describes the doctrine of *"the survival of the fittest"* in his book *The Selfish Gene*.[72] The evolutionary mythology is permeated with violence, disease and agonizing death, spilling blood for hundreds of millions of years. Through the evolutionary prism, or prison, death and extinction reign over every life, species and kind, including humanity. It is important for every reflective Christian to ask: if evolution were true what would this mean in terms of who GOD and JESUS is? If evolution was not mythology, then GOD would have to be the Little Tinkerer who instigated death and suffering in our world.

From His Word we know that CHRIST was the Creator of all things: ***"For by Him all things were created: things in heaven and on earth, visible and invisible, whether thrones or powers or rulers or authorities; all things were created by Him and for Him." (Colossians 1:15-17)*** To accept theistic evolution one has to believe that JESUS is either an absentee landlord who couldn't care less about His creation or that He intentionally or recklessly evolved all of the diseases and violence that afflict animals and people, past, present and future, for some cosmic scientific experiment or tragic comedy reality show. Under the theistic evolutionary big lab tent, JESUS would have to be the quintessential mad scientist and humans and every other creature would be nothing more than the lab mice or the "expendable crewman" of the show. Little Tinkerer would be the inventor of the suffering of physical death and the evolver of all of the misery of millions of years of disease and violence. The theistic god of evolution would be a demented gore addict that gets high on the agony of its creatures.

JESUS IS NOT A LITTLE TINKERER

In Revelation the angels and elders surround the throne of GOD praising GOD and proclaiming: *"You are worthy, our LORD and GOD, to receive glory and honor and power, for You created all things, and by Your will they were created and have their being." (Revelation 4:11)*

If the evolutionary fantasy were true, GOD would be the one who created or initiated all of the mutations, all of the genetic imperfection and weaknesses, all of the disease organisms, all of the "kill or be killed" hunger driven instincts that consume many creatures in our world today. This is not the JESUS or GOD of historical Biblical Christianity! Little Tinkerer, the god of theistic evolution, is a false god, an idol for those who reject the Holy loving GOD of the Christian faith.

If JESUS was the mad scientist behind the evolutionary process, that would make Him the greatest hypocrite who has ever lived, instead of the Holy SON of GOD. Coming into the world and pretending to be compassionate by healing a few hundred people during three short years of ministry, but at the same time being the instigator of all of the diseases that afflicted tens of billions of people and other life forms for millions of years, would have been hypocrisy higher than the heavens. The evolutionary scenario is an imperfect painful process. JESUS' fingerprints would not be found anywhere near the evolutionary crime scene of our world.

As we shared earlier, it is written: *"The LORD is good to all; He has compassion on all He has made." (Psalm 145:9)* Or, later in the same chapter: *"You open your hand and satisfy the desires of every living thing. The LORD is righteous in all His ways and loving toward all He has made." (Psalm 145:16,17)* There is no way that these verses could be true of the Little Tinkerer god of evolution. How is it compassionate or loving to evolve diseases and violence that cause the creatures you brought into being to suffer? Is GOD a sadistic animal abuser and/or a gleeful spectator of this suffering for millions and millions of years? This is the kind of god that theistic evolution conjures up, but it has nothing to do with the Living GOD, the FATHER of our LORD JESUS CHRIST, and our FATHER.

It is impossible that JESUS could be the author of the evolutionary doctrine of "the survival of the fittest". The revelation of GOD is that He is concerned about each individual, and has a special concern for the poor, needy, and helpless who cry out to Him. *"The survival of the fittest"* is the exact opposite of what JESUS said in His Sermon from the Mount: *"Blessed are the poor in spirit, for theirs is the Kingdom of Heaven...Blessed are the meek, for they will inherit the earth." (Matthew 5: 3-5)* The evolutionary doctrine of natural selection in the battle for existence is completely contradictory to what GOD reveals throughout His Word. According to GOD'S Word it is not the powerful and arrogant who will be blessed and inherit the earth, but those who are weak and humble. Those who recognize their need of GOD'S help and salvation, and cry out to Him for forgiveness and mercy, will be the blessed ones.

It is impossible that CHRIST could have looked at our world filled with pain, suffering, and death for hundreds of millions of years and said that it was very good, and that He was pleased with it, as He says in Genesis 1. Any god that had evolved all of this suffering and death wouldn't be worthy of the praise and honour of people. For which diseases do theistic evolutionists praise their Little Tinkerer for: cancer, diabetes, Parkinson's, or the worms that feed in the stomachs of malnourished children? Each individual is certainly free to choose to worship and serve the theistic evolutionary god if they want to, but Little Tinkerer is not the GOD of Abraham or Paul. GOD, as revealed by His SON and His SPIRIT in His Word is loving, holy, all-wise, all-powerful, perfectly just, and the source of everything which is good and right. Little Tinkerer would be a very limited, weak, scientifically challenged, and morally bankrupt being; a supreme mad-scientist with a macabre sense of humour.

The reason that many of the mainline protestant churches which have adopted theistic evolution into the fabric of their thinking are losing more and more members is simply because Little Tinkerer isn't worth worshipping or getting excited about. It is easy to understand why people wouldn't have any desire to worship that kind of a twisted god.

IS DEATH PHYSICAL OR SPIRITUAL IN GENESIS 2 - 3?

In Genesis 2 and 3 GOD declares to Adam and Eve that because of their sin, their disobedience to His command, they will surely die. Many theistic evolutionary believers have cornered themselves into arguing that the death GOD speaks about here is spiritual and not physical. Let's dig a little deeper and examine what this would mean.

People who believe that the death GOD speaks about in Genesis 2 and 3 is the death of the soul and not the body, logically must also believe that physical agony and death are good because their god created these from the beginning. Theistic evolutionary believers don't openly profess this truth about their version of god because they realize that it is quicksand beneath their feet. Nevertheless, it is obvious to all unbiased observers that their Little Tinkerer must delight in suffering, pain, disease, violence and death since under the evolutionary scenario these realities were present from the beginning, when GOD declared in Genesis 1 that everything was created good.

In any objective reading in Genesis 2-3 it is undeniable that death is a reference to physical death. Death is a direct consequence of humanity's sin, and completely contrary to GOD'S original creation. In Genesis 3, GOD says to Adam and Eve **"By the sweat of your brow you will eat your food until you return to the ground, since from it you were taken; for dust you are and to dust you will return." (Genesis 3: 19)** Is it the human soul or our body that returns to dust? The human spirit comes from the breath of GOD, not from the dust of the ground, and it returns to GOD, not to the dust. As it is written: **"The Spirit of God has made me; the breath of the Almighty gives me life." (Job 33:4)** **"Remember him—before the silver cord is severed...and the dust returns to the ground it came from, and the spirit returns to God who gave it." (Ecclesiastes 12:6,7)** Dust is the material element out of which our bodies were created. The judgment for Adam and Eve's sin was the physical death of their bodies and spiritual judgment to come.

The only avenue open for anyone who claims that the reference to death in Genesis 2 and 3 only applies to spiritual death and not physical death is to completely empty the words of their meaning by labelling them "mythology". In order to believe the evolutionary mythology of the Darwinian tree, they have to call GOD'S Word mythology. As we will see in chapter six, those willing to call Genesis mythology are also predisposed to calling the Gospels mythology. Genesis and the Gospels fit together like hand and glove. Evolution is a misfit that distorts both creation and salvation, both Genesis and the Gospels. It is not honouring to GOD to twist His Word by not accepting the clear intent that His Word communicates to us. We will examine this more closely in chapter 6.

WHAT ABOUT LASSIE AND FLIPPER?

Does the suffering and death that came into the world, as a result of Adam and Eve's sin, also include death for the other creatures that inhabit our planet? The answer of Scripture is "yes" it does. It is written:

"For the creation was subjected to frustration, not by its own choice, but by the will of the One who subjected it, in hope that the creation itself will be liberated from its bondage to decay and brought into the glorious freedom of the children of GOD." (Romans 8:20,21)

Humanity's sin has affected, and continues to affect all of the created order of our world and universe. Suffering and death affect all of the creatures of the earth, including the famous television animal heroes, Lassie and Flipper. It may seem unjust that humanity's sin should have that kind of consequence for all other creatures. Certainly there is much injustice in our world, but it is human injustice, and not the injustice of GOD. Our sin has far-ranging consequences beyond what we realize. It impacts everything in this world that was created for us. Our arrogant human tendency is to downplay the repercussions of our sin on our own lives, and on the created world around us. Sin's consequences, however, are very real and visible.

It is important to understand that GOD does not reveal everything to us in His Word. It is written: ***"The secret things belong to the LORD our God, but the things revealed belong to us and to our children forever, that we may follow all the words of this law."*** *(Deuteronomy 29:29)* The Bible, the Word of GOD in written form, is given to us humans and is focused on our relationship with GOD and with one another, past, present, and future. The relationship of GOD to the other creatures is not something that GOD reveals to us, except for the truth that He does care about them and it is His will that humanity look after and rule over them in a way which reflects GOD'S loving compassion. The revelation given us in Genesis 1 and 2 is that GOD created us to be His children and created the animals to be our pets. We are to name them, care for them, and rule over them. As a human race we have failed miserably in this area of stewardship, as we have in many others, and the animals suffer because of it. Humanity's injustice to creation will be rectified in GOD'S time.

Prior to the fall of the human race away from GOD, the Lassies and the Flippers and the other absolutely incredible creatures that surround us lived in perfect harmony with humans and with each other. It would have been a little bit like the symbiotic relationships that some creatures still have today, or like some Walt Disney movies. This is reflected in the statement that, prior to humanity's disobedience to GOD, all of the sea, land, and air creatures ate plants. It is written:

> *"I give you every seed-bearing plant on the face of the whole earth and every tree that has fruit with seed in it. They will be yours for food. And to all the*

beasts of the earth and all the birds of the air and all the creatures that move on the ground—everything that has the breath of life in it—I give every green plant for food." And it was so." (Genesis 1:29-30)

At this point in time GOD testified that animals thrived on eating plants and consequently, there was no need or desire for animals to survive from day to day by tearing each other apart for food. This harmony among all of the creatures is evidenced also in their complete submission to Adam, whom GOD put in charge. The dog-eat-dog, survival-of-the-fittest drive of the animal world developed as a result of the curse that human evil brought into the world. One day soon GOD will restore the original harmony that was present in the world at the beginning. As the prophet Isaiah wrote: *"The wolf will live with the lamb, the leopard will lie down with the goat, the calf and the lion and the yearling together; and a little child will lead them." (Isaiah 11:6*)

We praise GOD that in His time He is going to renew all of creation and remove the curse of human sin, liberating all of the created order from its bondage to decay, suffering and death, to once again thrive in a perfect environment.

ROSES ARE RED AND CHERRIES ARE DELICIOUS

As we consider the impact of human sin in bringing deterioration and death to our world, it is important that we not get confused by philosophical word games played by theistic evolutionists regarding cellular death or the death of microbes or plants. Their specious argument is that when Adam and Eve or the animals ate plants they were bringing death to those cells and, therefore they proclaim that death was in the world before the fall. Although plants and cellular life forms are living in a biological sense, they have no "nephesh" or breath of life in the Biblical sense. It is biologically accurate to speak of cellular death for these types of creations, but not death in the Biblical sense. Life forms such as plants, amoebas, bacteria, and viruses have no brain, no pain experience, and no sense that they are even alive. They were created for the precise purpose of giving our bodies all of the nutrients, vitamins, minerals, proteins and fats we need to thrive on, and, for anointing our world with beauty. They are software programs that create the beautiful and delicious delicacies that we can feast our eyes and appetites on to energize us. We can appreciate and praise GOD for the plants. They are amazing creations, as all plant scientists would acknowledge. Roses are red and cherries are delicious.

MIRROR, MIRROR ON THE WALL, WHO IS THE CULPRIT OF IT ALL?

"Mirror, mirror on the wall" who is the culprit of all of the suffering and death that permeates our world today? There are many who accuse GOD of being unjust and cruel for His allowing evil to intrude into our temporary world. To respond to this slander against GOD it is crucial to understand the basics of the HOLY SPIRIT'S teaching on suffering and death.

The universe GOD created was perfect in every way because that is who GOD is and what He does. A very important part of this perfection was not just the incredible design of everything, but the eternal nature of everything that GOD lovingly created. Death was not part of GOD'S amazing design, or part of His compassionate work. Even though we do not see that perfection today, by faith in GOD we know that in the beginning our planet was a perfect paradise very different from our deteriorating contemporary world.

Nothing that humanity does will be perfect or last for eternity. Everything we make will run down and eventually cease to function and exist. If we can imagine something that is perfect, then we are imagining something that is eternal. When something imperfect is introduced into the center of a perfect eternal system then the imperfection will affect the whole system, making the entire system imperfect. Once a perfect system has been infected at the center by an element that is imperfect, it will, from that time on, be subject to corruption. As more and more imperfection is injected into the center of the system it increases the rate of deterioration and eventually the whole system will collapse. This is the fate of our world because of human sin, Adam and Eve's sin, my sin, and yours.

JESUS said:

"If those days had not been cut short, no one would survive, but for the sake of the elect those days will be shortened...Immediately after the distress of those days: 'the sun will be darkened, and the moon will not give its light; the stars will fall from the sky, and the heavenly bodies will be shaken.' At that time the sign of the Son of Man will appear in the sky, and all the nations of the earth will mourn. They will see the Son of Man coming on the clouds of the sky, with power and great glory." (Matthew 24:22,29-30)

Note that CHRIST returns before everything collapses and is destroyed.

According to the revelation GOD has given us through His SON, His SPIRIT and His Word, human sin is the imperfection that entered the perfect universe and world that GOD created. The sun didn't immediately quit shining and the sky didn't suddenly collapse when Adam and Eve disobeyed the will of GOD, but their sin altered the perfection and harmony in the world so that a process of deterioration began in everything. From that moment on, their incredibly created bodies began to die, just as GOD had declared. As a result of their sin, the biological and chemical systems of all material life were infected, the foundations of the earth began to crumble, and the sun and the entire cosmos began to run down. The decay began at the time of Adam and Eve's fall into sin and the amount of decay has increased as humanity's sin has increased. This is evidenced through all kinds of symptoms such as earthquakes, plagues, violence in nature, destructive weather, weeds, disease, environmental destruction and increasing birth abnormalities due to mutations

It may be tempting to point a finger at Adam and Eve and blame it all on them, but through the prophets, apostles, and through His own SON JESUS CHRIST, GOD tells us the truth that we are all guilty of sin. We are individually and corporately responsible for the deterioration and death on our blue planet and in our universe. Our sin has brought corruption to all of creation.

Despite the advances in medical knowledge and technological achievements, diseases will increase, plagues will spread, and the death toll will always be one hundred percent as humanity continues down the path away from GOD (Romans 5:12; 6:23). The foundations of the earth will increasingly tremble as the sin of humanity multiplies exponentially around the planet. Yes, even the sun will quit shining and the stars will one day fall because of the cumulative effect of our individual and corporate sin. In Revelation 11, GOD is praised for bringing His judgment on those who destroy the earth:

> *"We give thanks to you, Lord God Almighty, the One who is and who was, because you have taken your great power and have begun to reign. The nations were angry; and your wrath has come. The time*

has come for judging the dead, and for rewarding your servants the prophets and your saints and those who reverence your name, both small and great—and for destroying those who destroy the earth." (Revelation 11:17,18)

GOD'S judgment on our sin generally comes in the form of allowing us to be subjected to the consequences of our sinful choices of both omission and commission *(Romans 1:18-32, Galatians 6:6-8)*. The decay in our universe is simultaneously the result of humanity's corporate and individual sin and the judgment of our Creator. Human life was at the very center of GOD'S creation. The cosmos and our blue planet, which reflect the awesome glory of GOD, were created as a beautiful paradise for the human race. We thoroughly messed it up. Consequently, as GOD'S SPIRIT declares in Romans 8, the whole universe is groaning as a result of humanity's fall from perfection.

Can anyone prove scientifically that the sin of humanity is the ultimate cause of earthquakes, tsunamis, famines, hurricanes, plagues, and other destructive phenomena? The answer, of course, is no. We cannot go back to the time of creation, and it is impossible for science to deal with the spiritual realities of life, and the interconnectedness of the spiritual with the physical. However, we know this to be true because GOD was there and has clearly revealed these truths in His Word.

Our corrupted human nature seeks to rationalize away our sin by blaming GOD, turning Him into a Little Tinkerer responsible for the mess our world is in. GOD is not the Little Tinkerer—we are. We keep going our own way and destroying the world and the life GOD has given. As in Genesis 3, we want to be our own little gods, doing everything our own way. Frank Sinatra sang: *"I did it my way."* Indeed, he did, and everyone did, with the exception of

CHRIST, who did it GOD'S way. How do you like the world run the human way? It is a long way down the path towards hell. Instead of blaming GOD, we all need to take an honest look into the mirror, repent and turn back to GOD and His way. It may be too late for our world, but it isn't too late for individuals and families. Today is still a day of salvation for many people.

WHY DOES GOD TOLERATE EVIL IN OUR WORLD?

The accumulating deterioration in our world has saturated our planet with suffering, violence, and death. Increasing personal, local, national, or international tragedies such as earthquakes, hurricanes, famines, plagues, wars, accidents, acts of violence and abuse have resulted in an explosion of anger towards GOD and towards others. There are a lot of people who blame GOD, accusing Him of being uncaring, unloving, unjust, and blatantly cruel in tolerating these evils.

For many it is a coin toss in which GOD loses either way. On the one hand, they continually berate GOD and the church for being intolerant, and on the other hand accuse Him of tolerating too much evil in our world. Having minds consumed with arrogance and anger, they are not going to give GOD a fair hearing. They conduct a kangaroo court in their own imaginations in which they have rehearsed the case against GOD and proclaimed their guilty verdict. In accusing and condemning the LORD our GOD, our Creator, as guilty of tolerating evil and not caring about the suffering on our planet, most do not take the time to listen or read what GOD says about evil or give GOD the opportunity to speak directly to the issues. Most of the hard core accusers have plugged their ears and closed their minds with self-righteous indignation, refusing to listen to GOD'S Words.

For those who are listening, GOD is speaking loud and clear. In *The Problem of Pain* C. S. Lewis wrote: "God whispers to us in our pleasures, speaks in our conscience, but shouts in our pains: it is His megaphone to rouse a deaf world."[73] GOD is not only speaking in a still small voice to those who are quietly listening to Him, but He is also shouting out to the deaf in our world. All over our world many are tuning in to listen to GOD'S voice. GOD says to you and me: *"The LORD Almighty is with us...Be still and know that I am GOD." (Psalm 46:7,10)*

The GOD of Abraham, Isaac, Jacob, Moses, and the FATHER of our LORD JESUS CHRIST is not an absentee landlord of our planet, nor is He silent about the reality of evil. In His message to us in His Word He has a great deal to say to us about evil's origin, its consequences, its destination, and His response to it. However, because much of what He says about evil is offensive to our self-righteous culture today, it is easier for many to condemn Him rather than listen to Him.

What is GOD saying? GOD says to us, as He said to His people of Israel, that we are not His judges, but He will judge us. It is written:

> *"Yet the house of Israel says, 'The way of the LORD is not just.' Are My ways unjust, O house of Israel? Is it not your ways that are unjust? "Therefore, O house of Israel, I will judge you, each one according to his ways, declares the Sovereign LORD. Repent! Turn away from all your offenses; then sin will not be your downfall. Rid yourselves of all the offenses you have committed, and get a new heart and a new spirit. Why will you die, O house of Israel? For I take no pleasure in the death of anyone, declares the Sovereign LORD. Repent and live!" (Ezekiel 18:29-32)*

In describing the human race GOD says:

"There is no one righteous, not even one...All have turned away. They have together become worthless... Their throats are open graves, their tongues practice deceit, the poison of vipers in on their lips, their mouths are full of cursing and bitterness. Their feet are quick to shed blood. Ruin and misery mark their ways, and the way of peace they do not know. There is no fear of GOD before their eyes...There is no difference for all have sinned and fallen short of the glory of GOD." (Romans 3: 9-23)

Our perspective on evil is usually very subjective. Anything that is seen as a threat or interferes with our own desires, plans or causes is seen as being evil. When others lie, steal, commit adultery with our husband or wife, or cut us off in traffic, we immediately label their actions as evil and respond with self-righteous indignation, and often anger. However, when we lie, steal, commit adultery, hurt others, or cut them off in traffic, we immediately rationalize our behaviour or words. We are all good at excusing our own sin and evil, but quick to point our fingers at the sin in others, just as JESUS said in Matthew 7.

We are rightly upset with terrorists who use the argument that *"the end justifies the means",* but we rationalize our own behaviour using the same stream of logic. In Canada and the United States, we were very quick to condemn the evil actions of the terrorists responsible for the destruction of over three thousand human lives on 9-11. The condemnation is just, but, nevertheless, hypocritical. Every day, in Canada and the United States, four to five thousand innocent unborn children created by GOD in His image are being ruthlessly destroyed by abortion.

We like to believe that we are good, better than those "other people" who are responsible for the pain and suffering of others. The truth is that either by our action or inaction, we are all responsible in the eyes of GOD for the accumulating evil and suffering in our world. It is written: **"There is no one righteous, not even one." (Romans 3:10)** There is no one who lives the perfect life of love and obedience to GOD, love for our neighbour, and faithful stewardship of His creation.

GOD'S standard of evil is objective and based on the whole truth, not just on a very selective and subjective view of the facts *(Romans 2:1-16)*. On the basis of His objective standard, GOD says that it is not just a few people here and there who are guilty of evil, but that the whole human race is a fallen, sinful, rebellious race, corporately responsible for the evil in our world. Nor does GOD allow us to excuse our evil on the basis of *"the devil made us do it"*. Like Adam and Eve in Genesis 3, we are not just innocent marionettes in the hands of an evil puppeteer. We are accountable to our Creator.

Contrary to the dogma of secular humanism, the human race is not basically good. Although the human race was good when GOD first created us, we have fallen from the glory of the living GOD. It is written: **"Their feet are quick to shed blood; ruin and misery mark their ways, and the way of peace they do not know." (Romans 3:15-17)** What an apt description of our human race. The evidence is all there in our personal and world history and it is clear beyond the shadow of any reasonable doubt that we are all guilty. The rabble-rousers may complain that GOD is too tolerant of evil but they fail to recognize the evil in their own hearts and lives and GOD'S amazing patience with all of us. If GOD did not tolerate any evil in our world, then He would have to destroy the whole human race immediately. In His infinite wisdom He has chosen not to do that yet. However, the Day of Judgment on our world is coming quickly. His patience with our world has a limit.

THE LIMITS OF GOD'S TOLERANCE

A second critical truth about evil that GOD makes very clear, from the beginning to the end of His Word, is that He will not tolerate evil forever. GOD'S love both conceives and limits His tolerance of evil. Tolerance has become a value in western Judeo-Christian societies but it is largely a misunderstood value. No one, except for a pure anarchist, would suggest that all evil should be tolerated at all times. Every civilization tolerates some actions and words but not others. Since we are all sinners, as both our consciences and GOD'S HOLY SPIRIT testify, tolerance in this world toward each other is a necessary and important characteristic up to a certain point. Tolerance beyond that point becomes destructive of individuals and the community.

In our present age GOD, in His love and sovereignty, tolerates sin and evil and their consequences of suffering, violence, and death in our world. In His Word GOD reveals that He is "long suffering" and "slow to anger". In extreme patience with our fallen world, GOD tolerates evil for a limited period of time until the complete number of those who will love and trust in Him are saved from their sin and brought into His glorious Kingdom through CHRIST. GOD promises through CHRIST and the SPIRIT that there is a day of justice and accountability and judgment coming for our whole world. For all who reject GOD and His Word and live a life of unrepentant evil, there will be an eternity of condemnation. Until that Day of Judgment, GOD calls His people to live by faith, trusting that His Word and promises will be fulfilled in due time (*Habakkuk 2*).

ALTERNATIVES TO TOLERANCE

Here are four theoretical alternatives to tolerating sin and its consequences in our world. As you will see none of the alternatives are good.

Option 1: The LORD GOD, in His sovereignty, could have elected never to create our universe or world in the first place. This would have been a classic non-starter.

Option 2: The second option would have been the immediate and complete extinction of the human race at the time of the first sin. This would have instantly eliminated Adam and Eve and all of their children, including us. If you are glad to be alive, praise GOD that He did not choose this second option.

Option 3: The third option would have been the removal of the freedom to choose between good and evil, transforming us into the lab mice in Little Tinkerer's cosmic experiment. The perfection GOD created involved making people in His own image, which includes the freedom to choose between good and evil, right and wrong.

Option 4: The fourth possible option to His limited and temporary tolerance of sin and its consequences would be for GOD to have created beings whose choices, actions, words, and thoughts would not have any consequences. If there were no consequences for our choices, there wouldn't really be any options to choose between, since they would all lead to the same non-consequence. This would be a ghostly nonsensical realm, a haunt for spirits disconnected from reality; a joyless, hopeless, barren world. Since GOD is Holy, this fourth option is not possible, since sin would always separate people from GOD.

For GOD, in His infinite wisdom, there probably were other options besides these four that are mentioned. The bottom line, however, is that everyone who trusts GOD'S infinite wisdom and love knows by faith that GOD'S decision to tolerate sin and its consequences for a few years, while He works at redeeming as many lives as possible, was the perfectly righteous and loving choice to make. In His complete foreknowledge *(Acts 2:23, Romans 8:28-30, Romans 11:2, 1 Peter 1:2)* and for the sake of His children who trust in Him, GOD chose to create our world and to send CHRIST to save us. We praise GOD that in His awesome wisdom and love He created the world, despite the evil that He knew would materialize and the sorrow that it would bring Him. He did this so that those who would believe and trust in Him could spend eternity with Him in His glorious Kingdom. We will praise Him for all eternity for His loving decision and salvation which will be completely revealed to us in the New Heaven and New Earth.

<u>GOD'S tolerance of sin and suffering and death for a few thousand years, is a completely different scenario than Little Tinkerer creating suffering, violence and death for a scientific experiment or cosmic joke lasting for a few billion years. We praise GOD for sending His SON to redeem us and His HOLY SPIRIT to convict, lead, teach, rebuke, correct, comfort and encourage His children during our pilgrimage in this world. Praise the LORD forever and ever!</u>

If GOD were to end all evil today, as He will one day soon, it would mean an end to human history in this world: *"fini", "fertig", "all over but the paper work"*: the contents of the Book of Life *(Revelation 20:12)*. It is coming soon. *It is written:* **"The LORD is not slow in keeping His promise, as some understand slowness. He is patient with you, not wanting anyone to perish, but everyone to come to repentance." *(2 Peter 3:9)*** The only path to restoration and redemption is the way of repentance and faith in the living GOD and His SON JESUS CHRIST. It is GOD'S choice to love each of us and offer us His wonderful amazing salvation through His SON. Each of us needs to receive His forgiveness and salvation.

PART B: Can GOD'S Love and Hell Co-exist?

COULD A GOD OF LOVE SEND ANYONE TO HELL?

Before leaving this chapter on the life-transforming truth that GOD is love and the giver of life, we need to turn our attention to one last crucial area linked to GOD'S love, namely the reality of eternal death in hell and eternal life in Heaven. One of the accusations so confidently and repetitively hurled against GOD and His people as proof that GOD is not love is that *"a loving GOD couldn't or wouldn't send anyone to hell"*. This accusation, thrown against GOD by both the world and the liberal church, customarily comes from the same lips as those who profess their abiding faith in evolution. Do they recognize how inconsistent they are? As we saw earlier in this chapter, the Little Tinkerer god they believe in would be the creator of thousands of millenniums of suffering in the world and offers nothing but eternal death to all people. Their accusation against GOD with respect to hell is loaded with disingenuousness. This is another "heads or tails, GOD loses" statement made by those who do not know or love GOD and do not believe His Word. Their thesis is that if GOD really is loving as GOD'S SON and SPIRIT have testified in His Word, then He could not send anyone to a place of eternal suffering. On the other hand, if GOD does punish people eternally, they say that He cannot be loving and therefore the whole of the Christian faith collapses. An unloving god isn't worth loving, obeying or serving. Either way they believe that they have put GOD and Biblical evangelical faith into checkmate. Nothing could be further from the truth.

GOD is love and hell is real. Hell is just and hell is good. The reality that many choose to follow the road to hell is not good (Matthew 7), but the fact that GOD created a place called hell is good. Everything that GOD does is holy and perfect, including the creation of a place called hell. If you are a believer in CHRIST, but the reality of hell has always made you very uncomfortable and maybe even ashamed, I hope that you will listen to the truth that GOD reveals to us about hell, as opposed to the false assumptions that so many make concerning it. As we will see, JESUS was completely upfront with people about the reality of hell and, as people who follow CHRIST we ought to be up front as well.

WHAT IS HELL?

What does GOD reveal to us about hell through His SON and HOLY SPIRIT in His Word? The purpose of this section is not to go into a detailed examination of all of the passages in the Bible that are related to the place that is commonly called hell. That would require a book all by itself. Rather we will take a look at the basic teaching concerning the reality of hell as the place of eternal condemnation and torment for those who choose to do evil without repentance, and without choosing to love and trust in GOD.

Here are some basic passages where CHRIST and the HOLY SPIRIT testify to the reality of hell:

1. *"Your hand will lay hold on all Your enemies; Your right hand will seize Your foes. At the time of Your appearing You will make them like a fiery furnace. In His wrath the LORD will swallow them up, and His fire will consume them." (Psalm 21:8,9)*

In this messianic psalm, proclaiming CHRIST'S victory over evil, GOD'S wrath is described as being like a fiery furnace in which unrepentant evil doers are consumed. This Psalm immediately precedes another Messianic Psalm which describes CHRIST'S suffering, death, and resurrection and the proclamation of His salvation to the ends of the earth. Before He returns to bring His judgement all of these other events had to be fulfilled with the result that many people will be brought into salvation. The last event to be fulfilled before He returns is the mission of bringing the message of salvation—forgiveness and eternal life through CHRIST to all the nations of the world (Matthew 24:14). This should spur all of us on to love and serve GOD and our neighbour, by bringing the Good News of CHRIST to all people before CHRIST returns and the Day of Judgment takes place.

2. *"But I tell you that anyone who is angry with his brother will be subject to judgment. Again, anyone who says to his brother, 'Raca, ' is answerable to the Sanhedrin. But anyone who says, 'You fool!' will be in danger of the fire of hell." "If your right eye causes you to sin, gouge it out and throw it away. It is better for you to lose one part of your body than for your whole body to be thrown into hell." (Matthew 5:21-23, 28-30)*

In these verses JESUS is giving a clear warning to everyone concerning the danger of hell. Like many other basic teachings of our Christian faith, the reality of hell is present already in the Old Testament, but it is only through the teaching of CHRIST that we are given a greater revelation of what hell is actually like. As Augustine of Hippo, an early Christian leader, put it: *"The New Testament is in the Old Testament concealed, the Old Testament is in the New Testament revealed."*

Theologians and others who claim to believe in CHRIST, but deny the reality of hell, obviously do not really trust CHRIST. It is superficial for people to say that they believe in CHRIST as GOD'S SON, but do not believe that what He taught is true. It is only through the teaching of CHRIST that the reality of hell becomes much clearer than it was in the Old Testament. The Christian Church received its teaching on hell from CHRIST. One cannot separate CHRIST from His clear teaching on hell.

The actual Greek word that is translated as "hell" is "gehenna". It is believed by many scholars that Gehenna derives its meaning from the Hebrew "ge hinnom", which is a reference to the Valley of Hinnom where, centuries before, the people of Israel sacrificed some of their children by fire to the Ammonite idol Moloch. This Valley of Hinnom became a dump where the garbage was burned. This is the picture JESUS used to illustrate the reality of hell. It is a place where the corrupted or morally and spiritually unfaithful are thrown out of GOD'S presence to be consumed by the fire outside.

When you are aware of a danger of which others may not be aware, it is a most loving, though sometimes difficult, task to warn them about the peril and how to avoid it. They may not heed the warning but that decision is their responsibility and not yours. This is precisely what JESUS did when He spoke about hell on many occasions through His ministry as recorded in the four Gospels.

Note also that JESUS never intended anyone to actually gouge out his or her eye, as some unbelievers mockingly suggest. His point, which believers have understood from the beginning, is that sin is serious and leads to the judgment of hell. Recently there was a news report of a man who had to cut off his own arm, which had become entangled in an industrial machine, in order to save his physical life. JESUS is saying that eternal life is worth much more than an eye, an arm or a leg. Therefore we ought to be on our guard against all kinds of sin in our lives: our thoughts, words, and actions.

> 3. *"They will throw them into the fiery furnace, where there will be weeping and gnashing of teeth." (Matthew 13:42)*
>
> *"Then he will say to those on his left, 'Depart from me, you who are cursed, into the eternal fire prepared for the devil and his angels... Then they will go away to eternal punishment, but the righteous to eternal life." (Matthew 25:41,46)*

In these two passages JESUS is speaking of the final judgement when GOD'S holy angels throw all those who do unrepentant evil into the eternal fire. Hell was originally prepared for the devil and his conspiring angels. In the eternal fire there is suffering of body and soul.

> 4. *"God did not spare angels when they sinned, but sent them to hell, putting them into gloomy dungeons to be held for judgment". (2 Peter 2:4)*

> "And the devil, who deceived them, was thrown into the lake of burning sulfur, where the beast and the false prophet had been thrown. They will be tormented day and night for ever and ever." (Revelation 20:10)

GOD has already pronounced judgment on the angels who joined satan in his rebellion, condemning them to eternal hell. At least some of them are being held in dungeons until the day they will be condemned to the lake of fire, where they will experience eternal suffering, together with satan, the beast or antichrist and the false prophet of Revelation 13.

5. *"Then death and Hades were thrown into the lake of fire. The lake of fire is the second death. If anyone's name was not found written in the book of life, he was thrown into the lake of fire."(Revelation 20: 14,15)*

   "But the cowardly, the unbelieving, the vile, the murderers, the sexually immoral, those who practice magic arts, the idolaters and all liars—their place will be in the fiery lake of burning sulfur. This is the second death." (Revelation 21:8)

"Sheol" in the Old Testament and "Hades" in the New Testament are references to the place of the dead who die apart from faith in GOD. The souls of the dead are held in Hades, which is like a temporary jail and place of punishment, until the final Day of Judgment when they will be publically convicted of their sin and evil and condemned to hell, the lake of fire. The lake of fire is here described as the second death. There are some who interpret these references to hell, the lake of fire, as being the "second death", meaning the final death or extinction of body and soul. Along the same line JESUS said: *"Do not be afraid of those who kill the body but cannot kill the soul. Rather, be afraid of the One who can destroy both soul and body in hell."(Matthew 10:28)*

> **6.** *"Outside are the dogs, those who practice magic arts, the sexually immoral, the murderers, the idolaters and everyone who loves and practices falsehood." (Revelation 22:15)*
>
> *"GOD is just: He will pay back trouble to those who trouble you and give relief to you who are troubled, and to us as well. This will happen when the LORD JESUS is revealed from heaven in blazing fire with His powerful angels. He will punish those who do not know GOD and do not obey the gospel of our LORD JESUS. They will be punished with everlasting destruction and <u>shut out from the presence of the Lord</u> and from the majesty of his power on the day He comes to be glorified in His holy people and to be marvelled at among all those who have believed." (2 Thessalonians 1:6-10)*

Revelation 21 and 22 are descriptions, both symbolic and real, of the eternal paradise CHRIST is creating for His people. In the last chapter of the Bible there is only one verse relating to hell. In that verse GOD clearly indicates the reality of the separation between Heaven and Hell. GOD'S people are with Him in His holy and eternal realm or city. Those who are unrepentantly evil are outside of this city, apart from GOD and His people. Those who choose to live without the loving plan and purposes of GOD are choosing to live apart from GOD in eternity, outside of His eternal presence and governance.

As the saying goes, *"Watch out what you wish for."* GOD grants the wish of those who want to live without His presence and the blessings of His love and salvation. Since GOD is the source of everything good, of life, love, hope and peace, apart from Him there is only agony and eternal death.

There are many truths which GOD has revealed to us concerning the reality of heaven and hell, but there are also many things that we do not fully comprehend. Heaven and hell involve completely different dimensions to life than what we are familiar with. Unless GOD has revealed a truth very clearly in His Word, it is best not to make up our own interpretation but to allow GOD'S Word to stand on its own. Those who are called to proclaim the Gospel of CHRIST and the Word of GOD have to be very careful not to add or subtract from the revelation GOD has given us in His Word. (Deuteronomy 4:2; 12:32; Proverbs 30:6; Revelation 22:18,19)

This much is clear in the Scriptures: hell is real, it involves great suffering and loss, and it is preceded by a judgment of condemnation from the Creator and Source of every good. The details are all in GOD'S hands, since He alone can judge and condemn. Since GOD is who He is we know that the details surrounding hell will all be in complete conformity with His infinite wisdom, power, holiness, justice and love. It is not given to us to fully comprehend all that hell entails, but it is given to us to know that GOD IS and will bring everything into judgment.

Our responsibility, following the example of CHRIST, is to proclaim from the rooftops everything that He has taught us *(Matthew 28:18-20)*, which includes a very clear and continual warning to all people concerning the danger of hell. Of course this is not our primary message to our world, but it is a crucial truth which cannot be deleted. Our central message is to present the Gospel or good news of JESUS CHRIST and His offer of salvation to eternal life through His own shed body and blood. CHRIST'S core message was always the good news of GOD'S eternal Kingdom where He will live with His people. GOD'S heart is to save people by His grace through the cross of CHRIST.

One of the central misunderstandings of many people, inside and outside of the Christian Church, is that GOD condemns people to hell apart from their will. This isn't precisely the case. Dr. Norman Geisler in his book co-edited with Ravi Zacharias entitled *Who Made GOD?* expresses this truth in his chapter on *"Tough Questions About GOD".* He writes:

HOW CAN A GOOD GOD SEND PEOPLE TO HELL?

"This question assumes that GOD sends people to hell against their will. But this is not the case. GOD desires everyone to be saved (see 2 Peter 3:9). Those who are not saved do not will to be saved. JESUS said, **"O Jerusalem, Jerusalem, you who kill the prophets and stone those sent to you, how often I have longed to gather your children together, as a hen gathers her chicks under her wings, but you were not willing" (Matthew 23:37).*

As C.S.Lewis put it: "There are only two kinds of people in the end: those who say to GOD, 'Thy will be done,' and those to whom GOD says, in the end, 'Thy will be done'. All that are in hell, choose it."...

Furthermore, heaven would be hell for those who are not fitted for it. For heaven is a place of constant praise and worship of GOD (Revelation 4-5). But for unbelievers who do not enjoy one hour of worship a week on earth, it would be hell to force them to do this forever in heaven! Hear Lewis again: 'I would pay any price to be able to say truthfully 'All will be saved.' But my reason retorts, 'Without their will, or with it?' If I say 'Without their will,' I at once perceive a contradiction; how can the supreme voluntary act of self-surrender be involuntary? If I say 'With their will,' my reason replies 'How if they will not give in?'"[74]

GOD'S judgment in condemning people to eternal death in hell is not contrary to their own decision. It is similar to salvation. All believers in CHRIST know that we are saved by GOD. We cannot in any way save ourselves; we can only be saved by GOD'S grace through CHRIST. It is written: **"For it is by grace you have been saved, through faith—and this not from yourselves, it is the gift of God—not by works, so that no one can boast." (Ephesians 2:8,9)** GOD saves us, but, at the same time GOD'S Word is clear and we understand that we need to choose to receive GOD'S offer of salvation through CHRIST and that GOD does not force anyone to trust in Him. By the power of His HOLY SPIRIT, through the testimony of the Gospel of CHRIST, GOD invites, calls, encourages, enlightens, and appeals to all people to repent of their sin and trust in His mercy and grace. Nevertheless, He does not turn us into robots that cannot refuse the offer. GOD chooses to save us and GOD'S SPIRIT calls us to choose to trust in Him. GOD gives us the gift of eternal life in Heaven and He invites us to receive this gift and live with Him in His eternal Kingdom.

Hell presents us with a similar situation. On the one hand, GOD condemns people to hell, but on the other hand they have chosen to live apart from and contrary to GOD'S will by refusing to repent and trust in CHRIST. Thus they choose an eternity apart from GOD. Based on their choice to live outside of His Kingdom, GOD sentences them to eternity apart from Him.

Those who reject GOD and do unrepentant evil will be publically judged and sentenced to the place called hell, which is also where they choose to be. The judgement on them will be by public trial during which all of their thoughts, words and actions will clearly reveal their unrepentant guilt and their choice of living in opposition to GOD and His will *(Revelation 20:12)*. Our Creator has the authority and the right to judge people according to His law and to save people by His grace. It is right and just that satan and his horde, including many people who have aligned themselves with satan's rebellion against GOD, will justly experience eternal separation from GOD and His salvation.

ARE PRISONS GOOD?

Is the creation of hell an action of love? Is hell good? Could a good GOD actually create a hell? Yes, yes, and yes. We can praise GOD that He created hell specifically for the devil and his conspiring angels. It isn't good that people will choose eternity in hell, rather than in GOD'S Kingdom, but it is good that there is a hell for those who rebel against GOD.

Let's put this in a contemporary earthly context. Is it a loving choice for a community, state or nation to employ police and build prisons to restrain those who choose to harm their neighbours by stealing from, raping, abusing, or murdering them? If you had the opportunity to vote on a referendum that would eliminate all prisons, set all criminals free, fire all the police officers, and encourage everyone to live any way they wanted, would you vote for it? Not likely. All of us recognize the importance of jails and prisons.

Complete anarchy, where people do whatever is right in their own eyes, could only produce a nation of angst and violence and not order and peace. The simple truth is that if all people were set free of negative consequences for their evil actions, many more people would turn to violence to accomplish their own self-centered desires. We see this happening in many societies where law and order have collapsed, usually due to corruption and war. If the prisons were emptied and the police let go, or the police and courts themselves became lawless, our towns and cities would soon deteriorate to the survival of the fittest doctrine of evolution. This is what is beginning to happen in some Mexican towns where the drug lords are taking control.

It would be great to be able to live in a nation where all people would voluntarily obey the laws and would do no harm to their neighbour, but we all know that it doesn't work that way with the human race. Prisons, not gulags, are made by people who care about and want to protect their families and communities from criminals who live by their own selfish desires and the laws of the jungle. Prisons are created out of love for justice and the desire to protect life, liberty, and peace.

Prisons are good and necessary and so is hell. Hell is a necessary protection that enables peace and liberty and love to eternally thrive. If GOD did not create a hell for the devil and his corrupted angels, then they would bring eternal war and conflict to GOD'S Kingdom. If there was no hell for those who choose to rebel against GOD, there would be no heaven for those who choose to love and trust in GOD. Prisons are built to punish the guilty and to protect the innocent from those who would harm them. Hell exists to punish the guilty and to protect GOD'S eternal family from those who desire to harm and destroy. GOD, in His infinite wisdom, created a hell for the devil and his angels. All those who choose to join satan in rebellion against GOD should not be surprised when they join satan in sharing his cell called hell.

There is a little verse or chorus that says:

"The devil is a sly old fox, if I could catch him I'd put him in a box.
I'd lock the door and throw away the key
for all of the dirty tricks that he's played on me.
I am so glad that I'm a Christian, I am so glad that I'm a Christian,
I am so glad that I'm a Christian, GOD has done so much for me."

If I could lock satan away forever, I would. I can't, but GOD can and will lock satan and all who do unrepentant evil away where they will never be able to trouble or harm any of GOD'S people again. We do not have to apologize to anyone for the reality of hell. However, we should weep for those who are on the path of sin that leads to hell. Love and truth should motivate us to bringing the Good News of CHRIST to them so that at least some will be "snatched from the fire" (Jude 23) May it be so!

HALLELUJAH! HEAVEN IS AROUND THE CORNER

The day is coming when satan and those who join his gang will get their just deserts; they will reap what they have sown. However, right now we are living in the days when it is written: *"Then the dragon was enraged at the woman and went off to make war against the rest of her offspring—those who obey God's commandments and hold to the testimony of Jesus." (Revelation 12:17)* and, *"This calls for patient endurance and faithfulness on the part of the saints."(Revelation 13:9)* Persecution of those who love and serve CHRIST is increasing around our world, including the western world. Ministries that are focused on helping persecuted Christians estimate that there are around 100,000-200,000 Christians every year who face martyrdom for their faith in CHRIST. Many more are imprisoned and tortured. Although this may not be an easy time to be a Christian, CHRIST has won the victory for us and satan's day of reckoning is soon coming. In the meantime we need to "keep on keeping on", spreading the message of CHRIST and His salvation and coming Kingdom. JESUS said: **"but he who stands firm to the end will be saved. And this Gospel of the Kingdom will be preached in the whole world as a testimony to all nations and then the end will come."** (Matthew 24:13-14)

> *"Then I heard what sounded like a great multitude, like the roar of rushing waters and like loud peals of thunder, shouting:*
> *"Hallelujah! For our LORD GOD Almighty reigns. Let us rejoice and be glad and give Him glory! For the wedding of the Lamb has come, and His bride has made herself ready. (Revelation 19:6-7)*

Hallelujah! Hallelujah! Hallelujah! Hallelujah! Revelation 19 proclaims the return of CHRIST and the complete destruction of the antichrist and his world system. In these verses the angels and the people of GOD surround the throne of GOD praising GOD and saying Hallelujah four times. The last Hallelujah focuses on GOD'S Kingdom which is coming.

If hell were the only eternal reality there would be nothing to celebrate. However, GOD- FATHER and SON and HOLY SPIRIT are bringing all who trust in Him into an eternal paradise. There is a new Heaven and Earth which JESUS has gone ahead to prepare for those who love Him, where everyone can spend eternity in the presence of our Creator and Saviour GOD.It is written:

*"You have made known to me the path of life;
you will fill me with joy in your presence,
with eternal pleasures at your right hand." (Psalm 16:11)*

"No eye has seen, no ear has heard, no mind has conceived what God has prepared for those who love him"—but God has revealed it to us by his Spirit." (1 Corinthians 2:9-10)

"For GOD so loved the world that He gave His one and only SON, that whoever believes in Him will not perish but have everlasting life. For GOD did not send His SON into the world to condemn the world, but to save the world through Him. Whoever believes in Him is not condemned, but whoever does not believe is condemned already for he has not believed in the name of GOD'S one and only SON." (John 3:16-18)

*"And I heard a loud voice from the throne saying, "Now the dwelling of God is with men, and he will live with them. They will be his people, and God himself will be with them and be their God. 4He will wipe every tear from their eyes. There will be no more death or mourning or crying or pain, for the old order of things has passed away." He who was seated on the throne said, "I am making everything new!" Then he said, "Write this down, for these words are trustworthy and true."
(Revelation 21:3-5)*

THE WAY OUT AND UP

GOD says that despite all our sin and evil, He is a merciful, loving, and patient GOD. He didn't create our world to have it end in the despair of sin, death and hell. The central message of His revelation is that He has given us His SON JESUS CHRIST to save us from our sin, and to save us for the eternal joy of His Kingdom. GOD desires to save all people no matter what they may have thought, said, or done in their lives. Through His one and only SON JESUS CHRIST, His death and resurrection, GOD has provided for the forgiveness of all of our sin and the glorious hope of an eternity spent with Him.

GOD understands human suffering, for His SON endured the painful death by crucifixion at the hands of our sinful human race, people who were no different than we are today. JESUS CHRIST came to forgive us for our sin and to give us an eternity in His Kingdom where there is no more evil, no more violence, no more death, only eternal joy and peace. Through His HOLY SPIRIT and faithful servants at work in our world today, GOD is calling to people of every country, language, race, and cultural and religious background to repent of sin and evil, and to believe in Him and in His SON JESUS CHRIST who suffered and died to save us. GOD freely offers this salvation to all people.

It is written: ***"The LORD is not slow about keeping His promises, as some understand slowness. He is patient with you, not wanting any to perish, but everyone to come to repentance." (2 Peter 3:9)*** The LORD is loving, just, and merciful, and He is being very patient with our evil world. His patience and tolerance however, will soon come to an end, and His judgment will come! As JESUS proclaimed: ***"Repent for the Kingdom of Heaven is near!" (Matthew 4:17)***

If you are reading this, and don't yet have the assurance of GOD'S love, forgiveness and salvation through JESUS

CHRIST, I encourage you to confess your sins to GOD and place your faith in Him and His Word. GOD loves you and sent His Son to die for your sin, and through His resurrection to give you eternal life. His kingdom is soon going to be here; this age of tears and sorrows will soon be history.

"But thanks be to God! He gives us the victory through our Lord Jesus Christ." (1 Corinthians 15: 57)

Chapter 6

"GOD is not a man, that He should lie!"

"I tell you the truth..." JESUS, the CHRIST
(Matthew 5:18, 8:10, 10:23,42 etc.)

"*I tell you the truth"*, or in the King James Version, *"Verily, Verily I say unto you"*, is the second most common phrase JESUS used in His preaching and teaching, after *"the Kingdom of GOD"!* In the four Gospel accounts of CHRIST'S life, JESUS is quoted saying this over twenty times. Considering that in the Gospels GOD'S SPIRIT gives us only a brief summary of JESUS' words, it is probable that JESUS used this phrase in His ministry hundreds of times as He went from village to village to bring the Good News of the Kingdom of GOD. What is true of CHRIST in His ministry is also true of GOD'S SPIRIT in His Word. From the beginning to the end of the Bible, GOD tells us very simply, unmistakably, and repeatedly, that He is the GOD of truth. This is part of GOD'S revelation of Himself.

    The Apostle John, inspired by GOD'S SPIRIT, introduces JESUS to the world in the opening verses of his gospel, by testifying that JESUS *"came from the FATHER, full of grace*

and truth" (John 1:14). CHRIST came into our world not only filled with the amazing grace of GOD, but also with the powerful truth of GOD. Let's listen to a few verses where GOD'S SPIRIT testifies to the truth and trustworthiness of GOD and His Word:

1. ***"God is not a man, that He should lie, nor a son of man, that He should change His mind. Does He speak and then not act? Does He promise and not fulfill?"***
(Numbers 23:18-20)

From the beginning to the end of His Word, GOD declares that He isn't like people because, He doesn't lie.

2. ***"If what a prophet proclaims in the name of the LORD does not take place or come true, that is a message the LORD has not spoken. That prophet has spoken presumptuously. Do not be afraid of him." (Deuteronomy 18:22)***

A corollary of this is, if anyone who claims to speak for GOD proclaims what isn't in accordance with the historical testimony in GOD's Word, that person is speaking presumptuously, and other people should not give them their attention or support.

3. ***"For the Word of the LORD is right and true; He is faithful in all He does." (Psalm 33:4)***

If GOD was a liar He would not be faithful and His Word couldn't be trusted.

> 4. "Do not withhold your mercy from me, O LORD; may Your love and Your truth always protect me." (Psalm 40:11)

The only real protection any of us has is that GOD is the GOD of love and truth.

> 5. JESUS said: "Man does not live on bread alone, but on every word that comes from the mouth of GOD" (Matthew 4:4)
>
> Jesus answered, "I am the way and the truth and the life. No one comes to the Father except through me." (John 14:6)
>
> "But when He, the SPIRIT of Truth, comes, He will guide you into all truth." (John 16:13)
>
> "Sanctify them by the truth; Your Word is truth." (John 17:17)

If GOD'S Word wasn't true we would be living on a lie, when we live on His Word (1 Corinthians 15:12-19). JESUS says that He is the truth. GOD'S Word in the flesh of CHRIST and in the words of Scripture is truth! It is not a mixture of truth and falsehood. If the HOLY SPIRIT did not speak the truth in His Word, then we would have no way of knowing the truth about GOD. Either everything GOD says in His Word is true or it could all be false.

> 6. "Not at all! Let GOD be true, and every man a liar. As it is written: 'So that You may be proved right when You speak and prevail when You judge.'" (Romans 3:4)

If GOD were a liar He could not pronounce judgment on anyone.

> 7. *"To the angel of the church in Philadelphia write: These are the words of Him Who is holy and true, Who holds the key of David." (Revelation 3:7)*
>
> *"They called out in a loud voice, 'How long, Sovereign LORD, holy and true, until You judge the inhabitants of the earth and avenge our blood?'" (Revelation 6:10)*
>
> *"I saw heaven standing open and there before me was a white horse, whose rider is called Faithful and True." (Revelation 19:11)*
>
> *"The angel said to me, 'These words are trustworthy and true.'" (Revelation 22:6)*

GOD is worthy of praise because He is the GOD of truth. If these and many other passages in the Scriptures that testify that GOD is the GOD of truth were not perfectly true, then the whole of Scripture would be undermined. There would certainly be no reason to believe in CHRIST if GOD and His Word were not trustworthy. The whole of the Christian faith would be utterly undermined if GOD was a liar. This is precisely why GOD'S enemy attacks the truth of GOD'S Word so stridently and persistently in every way and in every generation. The evil one has led many away from faith by casting doubt on the clear testimony of GOD'S Word, just as he did in the garden of Eden. If GOD were not a GOD of truth, then He wouldn't be worthy of trust, and we would all be lost and headed for eternal death.

FAITH IS TRUSTING GOD'S WORD

"Now faith is being sure of what we hope for and certain of what we do not see. This is what the ancients were commended for. By faith we understand that the universe was formed at GOD'S command, so that what is seen was not made out of what was visible...Without faith it is impossible to please GOD, because anyone who comes to Him, must believe that He exists and that He rewards those who earnestly seek Him." (Hebrews 11:1-3,6)

What does it mean when people say they have faith in GOD and CHRIST, and that they believe GOD is their Creator and the Creator of all things in Heaven and earth, visible and invisible? The evil one seeks to confuse people about basic truths of life and takes pleasure in twisting people's understanding of what faith is. Many are deceived into thinking that to have faith in GOD is merely to agree that some kind of god exists out there somewhere, who had something to do with creating our universe. This is not faith as revealed by CHRIST and the HOLY SPIRIT in the Scriptures. It is written: *"You believe that there is one GOD. Good! Even the demons believe that—and shudder." (James 2:19)* Merely affirming the existence of GOD is not Biblical faith; even the evil spirits acknowledge GOD'S existence. For people to claim that they believe in GOD without seeking Him, listening to Him, trusting Him, or obeying Him is not the Christian understanding of faith. This false substitute for genuine faith deceives many people who call themselves Christians and even call JESUS "LORD" without ever trusting in GOD or following CHRIST. JESUS said:

"Not everyone who says to me, 'LORD, LORD,' will enter the kingdom of heaven, but only he who does

the will of My FATHER who is in heaven. Many will say to me on that day, 'LORD, LORD, did we not prophesy in Your Name, and in Your Name drive out demons and perform many miracles?' Then I will tell them plainly, 'I never knew you. Away from Me, you evildoers!'" (Matthew 7:21-23)

A rich young man came to JESUS one day and asked Him a crucial question concerning how a person can receive eternal life *(Matthew 19:16-26)*. JESUS answered that he should obey the commandments of GOD, sell his possessions and give the proceeds to the poor, and then he should follow CHRIST. The man left dejected choosing not to follow CHRIST. The young man certainly believed in GOD and even tried to live a law-abiding life, but he had no faith in CHRIST, no faith in GOD to obey GOD'S will for his life by following CHRIST. There are many people in churches today who are living dejected and fruitless lives because they are unwilling to trust GOD'S Word and to obey CHRIST.

Simply acknowledging the existence of GOD is not faith. Faith is trusting in GOD, trusting His Word – His teachings, His doctrines, His precepts, His commands, His promises, His encouragements, His salvation. Faith is trusting that what GOD says to us in His WORD by His HOLY SPIRIT and through His SON is true and that we can base our lives and eternity on it. Faith is being sure of what we hope for because GOD is trustworthy and faithful in fulfilling His Word and promises to those who trust in Him.

Faith can have a solid foundation only if GOD is a GOD of His Word, a GOD who doesn't lie or mislead those who trust in Him. Having faith in GOD would be senseless if He was not a GOD who kept His Word. The center of our Christian faith is a relationship of love and trust in GOD. Without love and trust a relationship cannot develop in a positive way. If GOD was not a GOD of His Word, then we could not trust that what He says in His Word is true and we could not have a love relationship with Him.

In many church circles these days we often hear people say that the doctrines, commands, and teachings of GOD including creation teachings, are not important because our Christian faith is all about a love relationship with GOD. It sounds enlightened and freeing, but it is a rationalization for not seeking GOD'S truth and acting on it. ***JESUS said: "If you love Me, you will keep my commandments!" (John 14:15)*** If the doctrines, commands, and teachings of GOD and CHRIST are not important, how do we know that GOD is love, or that heaven is real, or that CHRIST died on the cross for our sins? How do we know how we should love others if we don't know what GOD teaches us about how to love others in a responsible way? Love is an empty word that can be filled with all kinds of selfishness and evil if it is not filled with CHRIST'S SPIRIT and Biblical teachings.

Psalm 119, the longest chapter in the Scriptures, is found at the very center of the Bible. It is a love chapter praising GOD for His amazing Word and how blessed we can be living by His every Word. When satan sought to tempt CHRIST to disobey GOD'S will for Him, JESUS responded with the words: ***"Man does not live on bread alone, but on every Word that comes from the mouth of GOD" (Matthew 4:4)*** The Christian life is faith in GOD and in His revealed Word. Faith understands and knows that the universe was formed by GOD'S command during those six days of creation because that is what GOD has clearly revealed by His SPIRIT and SON *(Hebrews 11:1-2)*. If you believe that GOD is the GOD of truth and that His Word is truth, as JESUS taught, then you will want to read, study, meditate, and memorize His Word so that you can live by it every day.

JESUS AND ADAM

"For what I received I passed on to you as of first importance: that CHRIST died for our sins according to the Scriptures, that He was buried, that He was raised on the third day according to the Scriptures...Whether, then, it was I or they, this is what we preach, and this is what you believed... And if CHRIST has not been raised, our preaching is useless and so is your faith. More than that, we are then found to be false witnesses about GOD...And if CHRIST has not been raised, your faith is futile; you are still in your sins...If only for this life we have hope in CHRIST, we are to be pitied more than all men...For since death came through a man, the resurrection of the dead comes also through a man. For as in Adam all die, so in CHRIST all will be made alive." (1 Corinthians 15:3-22)

GOD'S Word is truth when talking about JESUS or Adam. Although the HOLY SPIRIT, speaking through the Apostle Paul, is writing specifically here about the resurrection hope that an individual receives through faith in CHRIST, the heart of these verses relates to the integrity of GOD in the Scriptures. Note that CHRIST'S death and resurrection are in fulfillment of the Scriptures and that death is connected directly to Adam's fall into sin, just as the resurrection hope is connected to CHRIST'S resurrection. To believe the HOLY SPIRIT'S testimony to CHRIST'S resurrection, while denying His testimony to GOD'S creation of the world and Adam and Eve's fall into sin and the shadow of death cannot be done with integrity for the SPIRIT binds the two together.

JESUS said: *"I have spoken to you of earthly things and you do not believe; how then will you believe if I speak of heavenly things" (John 3:12)* Why would people believe in the resurrection hope of eternal life with GOD in a new Heaven and earth if they did not believe what GOD has said about this world in which we are living today? If what GOD has spoken and written about creation in Genesis isn't reliable, then why would anyone believe anything else in the Bible? Thank and praise GOD every day that He is the GOD of truth. He does not lie, whether it is about this world or about heaven.

From the Gospels and other pages of history we know that CHRIST lived two thousand years ago. GOD'S SPIRIT has given us the basic genealogies for JESUS taking us back to Adam and Eve *(Luke 3:23-38)* about six thousand years ago. If one believes in JESUS, one needs to believe in Adam and Eve, for the HOLY SPIRIT testifies that JESUS is their descendant. Evolution has no place for Adam and Eve and no Garden of Eden for them to live in. If people do not believe what GOD'S SPIRIT has revealed concerning Adam and Eve, why would they believe what GOD'S Word says concerning their descendant CHRIST? Adam and Eve are historical people in the Bible just like JESUS is, except they sinned and CHRIST did not. Adam and Eve were just like us, and CHRIST came to save us all.

GOD'S truthful and trustworthy nature is foundational to every other truth about GOD. It is central to the truth of creation. From the first verses of Genesis to the end of Revelation, GOD has revealed that He is the Creator of Heaven and Earth and of all things in them. GOD has also said that people could know that He is GOD through what He has made. The central dogma of evolution is that all things have come into existence simply through chance and time according to the laws of nature without the necessity of any creator GOD. This is a complete rejection of what GOD reveals in Psalm 19 and an exact fulfillment of Romans 1. In Psalm 19 GOD says that we can see His glory through the creation that He has made. In Romans 1 GOD says that, because of unbelief, people will reject the clear testimony of creation to its Creator. As we saw in the first chapter GOD'S infinite wisdom and power are clearly revealed throughout creation for anyone with an open mind: a mind not enclosed in the little box of materialism.

SHOULD GENESIS BE TAKEN LITERALLY?

"And we have the word of the prophets made more certain, and you will do well to pay attention to it, as to a light shining in a dark place, until the day dawns and the morning star rises in your hearts. Above all, you must understand that no prophecy of Scripture came about by the prophet's own interpretation. For prophecy never had its origin in the will of man, but men spoke from God as they were carried along by the Holy Spirit." (2 Peter 1:19-21)

Most theistic evolutionists believe that Genesis, particularly the first few chapters, should be taken symbolically or mythically, and not literally. Those who believe that science cannot err argue that the Bible should not be taken "literally", anytime it teaches something contrary to their limited materialistic perspective. The word "literally" is purposefully misused by those who reject the historical testimony of the Scriptures. To accept something as literal means to accept it according to how it was intended to be read, the plain, simple logical interpretation of the words in their context. Many parts of Scripture are historical, others philosophical, others prophetic or poetic. There is no question that Genesis was intended as a book of history, and the parables of CHRIST were stories highlighting important teachings. Both are literally true, one as a historical witness, and the other as a historical witness to a story JESUS told to illustrate a point. JESUS' use of parables in the Gospels does not turn Genesis into a collection of parables. In the Scriptures, it is generally very clear whether a passage is historical narrative or a parable. Only those who want to deny certain historical accounts of GOD'S Word like to confuse the two.

It is enlightening to understand that those who openly reject the GOD of the Scriptures prefer to distort the message of GOD in the Bible by taking the historical literal truths of Scripture and treating them as mere myth or symbol. Because this is a simple way of confusing people who do not know GOD and His Word, it is a technique used by all false preachers and teachers. In His Word GOD repeatedly warns those who sincerely believe in Him to watch out for all these false teachings. GOD speaks both literally and symbolically in His Word, but only those who come to know and love GOD through faith and who submit to GOD'S HOLY SPIRIT can properly understand and apply His Word.

WHEN IS A DAY A DAY?

A common attack on the creation account in Genesis revolves around the word "yom" which is rightly translated "day". Since the word "yom" in Hebrew or "day" in English is on the rare occasion used in reference to an "age" or "era" of time, some theistic evolutionists suggest that it is used in this way in Genesis. In English today we may use the phrase "in JESUS' day" or "in Charlemagne's day" to refer to that period of human history in which they lived. There are a few instances in Scripture where "yom" or "day" is used in reference to a different time frame other than a regular day, but there is no indication anywhere in Scripture that the word is used in that way in the creation account.

Scripture is its own interpreter! Scripture interprets Scripture or, more precisely, the HOLY SPIRIT gives the proper interpretation of Scripture through the prophets, apostles and centrally through JESUS, the Word made flesh. Genesis 1 clearly speaks of creation in six normal days, and that is how all of the prophets and apostles understood it, because they all accepted the historical meaning of the words. When it comes to faith, it is wise to choose the simple straightforward interpretation of the words of Scripture, held by CHRIST and all of the faithful saints of the past who have trusted in Him, rather than the interpretation of any scholar or scientist who ignores the plain contextual meaning of the words.

The clear intended meaning of the word "yom" in Genesis 1 is in reference to the common Jewish understanding of a day, from one evening to the next. Note that after His creative activity on each of the six days GOD says: ***"and there was evening, and there was morning".*** There is no question that, from the beginning, GOD'S people have clearly understood that GOD'S creation of the cosmos, world, and life took place in six calendar days. For an excellent exegetical and historical analysis of the Hebrew word "yom" in Genesis 1 see Dr. Jonathon Sarfati's second chapter *"The Days of Creation"* in his book *Refuting Compromise,* where he meticulously lays out the clear reasons why "yom" can only reasonably be interpreted as a regular day.[75]

The problems are overwhelming for those who choose to believe that "yom" is used here in reference to huge amounts of time, with each day representing hundreds of millions, or even billions of years. Each day has periods of both light and darkness. If each day contained millions of years of darkness, all life created on earth that day would have perished during the long period of night. Plants, animals and people all have internal clocks based on the light and darkness of a regular day. Life could not survive during millions of years of darkness.

For those who believe the evolutionary interpretation of the fossil record, the order of the creation is very different from the order given in the creation days of Genesis. In Genesis 1 the plants are created on the third day, prior to the creation of the sun on the fourth day. This would be impossible according to the theory of both cosmic and biological evolution. You simply cannot have plants evolving millions of years before the sun begins to shine. According to Genesis the birds are created on the fifth day and land creatures are created on the sixth day. This is the reverse of the evolutionary scenario that teaches that birds evolved from the land creatures called dinosaurs.

The only alternative to believing that "yom" is used as a regular twenty-four hour day in Genesis 1 is to believe that all of the creation account and most of Genesis is mere mythology. People are free to believe that if they want to, but they should realize that their belief is contrary to that held by CHRIST, His prophets and apostles, and the saints over the ages who have upheld the historical nature of Genesis.

The historical witness of Genesis details GOD'S institution of marriage, referenced by CHRIST in Matthew 19, genealogies including CHRIST'S (Matthew 1, Luke 3), GOD'S judgment on the world of Noah's day, the covenant of faith between GOD and Abraham, and the truth that GOD acts miraculously beyond the presently known laws of the universe. The only way in which Genesis can be authentically read and interpreted is as history. There is never any hint anywhere in Scripture that Genesis was to be understood in any way other than as a historical book. Any metaphorical or mystical approach to Genesis is simply a cover for unbelief in this part of GOD'S Word.

Martin Luther, the sixteenth century reformer wrote:

"And whoever is so bold that he ventures to accuse GOD of fraud and deception in a single word and does so wilfully again and again after he has been warned and instructed once or twice will likewise certainly venture to accuse GOD of fraud and deception in all of His words. Therefore it is true, absolutely

and without exception, that everything is believed or nothing is believed. The HOLY GHOST does not suffer Himself to be separated or divided so that He should teach and cause to be believed one doctrine rightly and another falsely."[76]

Unbelief in one part of GOD'S Word is like a cancer that spreads to every other part that an individual chooses to disregard. The cancer of unbelief spreads from one part of GOD'S Word to another and from one individual or denomination to another, to all who do not hold firmly to GOD'S Word. Only the radiation of the love and truth of GOD'S Word in CHRIST can destroy the cancer of unbelief. If you are having trouble believing GOD'S Word on creation, then listen, read and meditate on all of the teaching of CHRIST and the HOLY SPIRIT, and His love and truth will set you free.

WAS JESUS DECEIVED AND A DECEIVER?

Was JESUS Himself deceived? Did He help to deceive others? It is necessary for those who hold to theistic evolution to believe that JESUS was just a man of His time who was Himself deceived concerning creation and helped to deceive others. It is interesting to note that in the context of liberal theology, when JESUS speaks or ministers to women, Samaritans, lepers, and the various other sinners that came to Him, He is seen as being avant-garde, freed from the patriarchal constraints of the Old Testament and of the culture of His day. However, when He speaks about the truth and authority of Scripture *(Matthew 4:4,7,10; 5:17-19; 26:54; Luke 24:25-47; John 5:39; John 10:35; John 13:18; John 17:12-17)*, creation and marriage *(Matthew 19:4-6)*, Noah *(Matthew 24:36-39)*, or Jonah *(Matthew 12:38-41; Luke 11:29-32)*, He is seen as being a captive to the unenlightened thinking of His generation. Liberal theology believes in itself, and not in JESUS and His Words.

It is clear from the teaching of CHRIST that He believed and accepted all of the Old Testament as literal truth which, in terms of Genesis, means historical truth. I don't know of any serious Biblical scholar who suggests that JESUS believed Genesis was mere mythology or parable. If JESUS were a mere man, we could ignore any of His teachings that are contrary to our contemporary myths. That option is not open for those who know CHRIST as the living GOD—GOD'S Word in the flesh. A disciple of CHRIST, a Christian, is to trust in and obey the teachings of CHRIST. JESUS believed in the historicity of Genesis and therefore all, who sincerely believe and trust in CHRIST, will also accept the Genesis revelation as historical. The only alternative is to believe that JESUS was deceived and helped to deceive others, in which case it would be nonsense to trust in Him. We either take JESUS at His Word or reject Him.

JESUS TESTIFIED TO THE FLOOD

> *"As it was in the days of Noah, so it will be at the coming of the Son of Man. For in the days before the flood, people were eating and drinking, marrying and giving in marriage, up to the day Noah entered the ark; and they knew nothing about what would happen until the flood came and took them all away. That is how it will be at the coming of the Son of Man." (Matthew 24:37-39)*

JESUS, who always professed that He told the truth, and nothing but the truth, testified to the flood of Noah's day. Most theistic evolutionists reject not only the Genesis account of creation, but also the account of a global flood. By rejecting the global flood they imply that their knowledge is superior to CHRIST'S. JESUS connects the historical truth of the global flood with the promise of His second coming to the world. Thus, to reject the flood is to deny the testimony of CHRIST to the flood, and to His return. This is reminiscent of the Gnostic teachers in the early church who led many away from CHRIST by claiming that their testimony to the "Jesus" they believed in was greater than the testimony of the Apostles to CHRIST. Those who reject the global flood are either rejecting the authority and testimony of JESUS to the flood, or are rejecting the testimony of the apostles to JESUS. Either way, they are on a very slippery slope heading away from the JESUS the HOLY SPIRIT reveals in the Scriptures.

THE FLOOD OF NOAH'S DAY WASHES AWAY THE EVOLUTIONARY MUCK

The global deluge of Noah's day not only cleansed the world of the filth of evil that had accumulated on the planet at that time, but it also washes away the evolutionary muck today. The evolutionary story not only denies the creation account in Genesis but also the account of the flood described in Genesis 6-9. The Biblical historical accounts of creation and of the flood are linked to each other, not only because of their close proximity in Genesis, but also because they are both completely opposed to the evolutionary myth. The vast majority of theistic evolutionists reject the account of the global flood because it simply does not fit with the evolutionary scenario of millions of years of geological time. The evolutionary long ages would transform the global flood into a global deception that GOD has played on His people for over four thousand years. To be sure, theistic evolutionists don't openly accuse GOD of lying in the flood account, but they interpret the flood account as a mythological story, a tall tale that grew out of a local flood.

The historical account of the flood in Genesis 6-9 undergirds the clear Biblical teaching of CHRIST and the prophets and apostles of a coming day of judgment for our world today. The vast majority of theistic evolutionists generally do not believe in or proclaim the reality of GOD'S judgment on our world, and therefore the global flood is an abomination to them. How could GOD judge humanity for sin and evil if He was the Little Tinkerer responsible for the enormous suffering and death in our world during their imaginary eons of time? However, in rejecting the account of the global flood they are exposing the truth that they choose to trust the interpretation of some scientists rather than the clear teaching of the Word of GOD.

Let's take a closer look at Genesis 6-9. There is no question that the flood account in Genesis 6-9 is written in Hebrew historical narrative form; there is no poetic story here. The flood account immediately follows the genealogy from Adam to Noah given in Genesis 5. Prior to the advent of the evolutionary myth it was universally accepted that these chapters spoke of historical events in the annals of the world. All faithful servants of GOD who have preceded our generation proclaimed the truth of the global flood. Although there are some legitimate, but very minor, issues involving the interpretation of the first four verses, the rest of the account is a very straightforward description of the cause and events of a global deluge. This cataclysm destroyed all terrestrial life on the planet with the exception of Noah, his family, and the creatures they had with them on the ark.

GOD'S words in these chapters could not be any clearer:

"The LORD saw how great man's wickedness on the earth had become, and that every inclination of the thoughts of his heart was only evil all the time. The LORD was grieved that He had made man on the

earth, and His heart was filled with pain. So the LORD said, 'I will wipe humanity, whom I have created, from the face of the earth—men and animals, and creatures that move along the ground, and the birds of the air—for I am grieved that I have made them.' But Noah found favour in the eyes of the LORD...Now the earth was corrupt in GOD'S sight and was full of violence. GOD saw how corrupt the earth had become, for all the people on earth had corrupted their ways. So GOD said to Noah, 'I am going to put an end to all people, for the earth is filled with violence because of them. I am surely going to destroy both them and the earth. So make for yourself an ark of cypress wood...I am going to bring floodwaters on the earth to destroy all life under the heavens, every creature that has breath of life in it. Everything on earth will perish. But I will establish My covenant with you...'" (Genesis 6:5-8,11-14,17,18)

GOD'S words are simple and unambiguous for Noah and for us. What actually happened?

"In the six hundredth year of Noah's life, on the seventeenth day of the second month—on that day all the springs of the great deep burst forth, and the floodgates of the heavens were opened. And rain fell on the earth forty days and forty nights...For forty days the flood kept coming on the earth, and as the waters increased they lifted the ark high above the earth. The waters rose and increased greatly on the earth, and the ark floated on the surface of the water. They rose greatly on the earth, and all the high mountains under the entire heavens were covered. The waters rose and covered the moun-

tains to a depth of more than twenty feet. Every living thing that moved on the earth perished—birds, livestock, wild animals, all the creatures that swarm over the earth, and all mankind. Everything on dry land that had the breath of life in its nostrils died. Every living thing on the face of the earth was wiped out; men and animals and the creatures that move along the ground and the birds of the air were wiped from the earth. Only Noah was left and those with him on the ark." (Genesis 7:11-12,17-23)

Notice the precise date is given for the day the flood began: the seventeenth day of the second month during Noah's six hundredth year of life. Dates, history, and genealogies go together; Genesis is loaded with them.

A century ago many laughed at the idea of huge bodies of water underneath the earth's surface. Today it is generally conceded that mega lakes and seas exist in the earth's crust and mantle. In fact, some scientists estimate that in the earth's crust and mantle there may be ten times the amount of water that there is on the surface. GOD had already revealed this back in Noah's generation, a long time before the advent of modern geology. Our planet is the only place in the universe confirmed to have liquid water, and we are surrounded by it on the surface and above and below it. Praise GOD for the gift of water.

GOD tells us that the waters rose more than twenty feet above than the highest mountain, completely submerging all terrestrial life. No local flood here. As many creation scientists have pointed out, why would GOD have commanded Noah to build an ark the size of a football field if all he had to do was move a few miles to higher ground. If the flood was merely local, Noah and his family could have walked out of the flood zone in a few days or weeks, rather than spending years building the ark. Why gather animals together when the local animals could have trotted off to the nearest hills? The rabbit and tortoise could have had a race to the closest hill.

It is hard to comprehend how anyone could read or hear the account of Genesis 6-9 and believe that it is describing a local flood. They can believe that if they want to but they should not call it Christianity, thereby confusing themselves and others. Those who don't believe in the flood don't believe in CHRIST'S testimony to the flood. Can you believe in CHRIST and not believe what He taught?

THE RAINBOW: THE SIGN OF GOD'S FAITHFULNESS OR DECEIT?

The rainbow is "the most incredible light show on earth" as someone has described it. Every day in various places throughout our world the rainbow testifies to GOD'S faithfulness to His promises.

"Then GOD said to Noah and to his sons with him:...I establish my covenant with you: Never again will all life be cut off by the waters of a flood; never again will there be a flood to destroy the earth...This is the sign of the covenant I am making between Me and you and every living creature with you, a covenant for all generations to come: I have set My rainbow in the clouds, and it will be a sign of the covenant between Me and the earth. Whenever I bring clouds over the earth and the rainbow appears in the clouds, I will remember My covenant between Me and you and all living creatures of every kind. Never again will the waters become a flood to destroy all life...This is the sign of the covenant I have established between Me and life on the earth." Genesis 9:8, 11-15,17

A covenant in modern terms would be an agreement or contract in which legal entities make certain commitments or promises. The rainbow is GOD'S signature to His covenant promise that He will never bring His judgment on our world again by way of a global flood. Now if the flood of Noah's day had been just a local flood then the rainbow would not be a sign of GOD'S faithfulness to His promises. If the flood had been local, the rainbow would be a sign announcing that GOD does not keep His promises, because every year many animals and people die in local floods around the world. During the summer of 2010 (while I am writing this) the news is often dominated by the incredible floods in Pakistan and China and many other parts of our world where thousands have died, and many more thousands are expected to die.

There are an estimated eighteen hundred thunderstorms occurring on our planet at any given time, about a million every year. Thousands of rainbows testify to people in every corner of our world either that GOD is a faithful GOD who keeps His promises or that He is an unfaithful liar who spreads myths to people. For the believing Christian, this is not just a matter of the historical events of Noah's day, but whether or not GOD can be trusted, whether His Word and covenants are reliable. If there was no global flood, then GOD'S Word and promises, including His promises through CHRIST, would be null and void. The rainbow represents the first recorded covenant that GOD made with people. If His first covenant is not reliable, then why trust the covenant through CHRIST? If the rainbow is not a sign of GOD'S faithfulness to His promises, neither is the cross. Scripture is woven together and cannot be torn apart without damaging it.

Praise GOD for the rainbow that shines in every corner of the planet as a clear testimony of GOD'S faithfulness to His Word! Praise GOD for the sign of the cross that His faithful people have lifted up all over our world so that people everywhere can know of forgiveness for their sin and the hope of eternal life!

ALL TRUTH IS GOD'S TRUTH

A number of times in the writings of theistic evolutionists the accusation is made that if evolution is not true then GOD would have deceived our world because "science" clearly supports evolution. They argue that since Christians believe that GOD is a GOD of truth it is necessary to believe in evolution because "science" says that evolution is true. Another way of saying this is: GOD is the GOD of all truth and all truth is GOD'S truth, and therefore the "truth" of evolution must also be GOD'S truth. Their argument doesn't hold any water because:

1) It doesn't comprehend the reality of the power and wisdom of GOD;
2) It assumes that whatever evolutionary materialistic science says is true; and,
3) It always limits GOD to the present state of scientific materialism.

It is important to remember again what JESUS said to the Sadducees, the religious liberals and materialists of that time: *"You are in error because you do not know the Scriptures or the power of GOD." (Matthew 22:29)* Quite frankly, as was stated earlier, the problem of theistic evolutionists is that the GOD they believe in is far too small. They have exaggerated human capabilities and minimized GOD'S power, by not trusting GOD'S ability to communicate His Word faithfully.

GOD never gives scientific explanations for any of His miracles. If GOD gave us a truly scientific explanation of how He made the universe or life, all the books in the world would not be able to hold the information, and it would all be beyond our present scientific understanding. In place of the scientific details, GOD in His wisdom gave us the historical version that literate people down through history could understand and believe and share with those who were illiterate, so that all could know the truth about CHRIST and His eternal Kingdom.

Any material entity that GOD creates out of nothing has the appearance of age, whether it is a rock or an animal. When Adam and Eve were created they naturally had the appearance of age. Was GOD deceiving us? Not at all; rather, appearances can be deceiving and inferences are often misleading. In the historical narrative of Genesis GOD tells us that these events were all miraculously accomplished by His power, wisdom and love in a very definite and short period of time.

It would be incredibly deceitful for GOD to tell us that He created the cosmos in six days, if the universe really evolved by itself over billions of years with His input being limited to igniting the spark and then quietly sitting in the bleachers. If that were the case, then for thousands of years God would have been intentionally deceiving His people who have trusted in Him. If in the past GOD has deceived His people who believed in Him, could you be sure He is not deceiving you about JESUS and eternal salvation?

It would be deceitful for GOD to tell us that He created human life in His image in a single day if humans are really the undirected result of a few billion years of mutations and natural selection in which He had minimal involvement. It would be deceitful for GOD to tell us that He created a very good universe if the history of our world from the beginning of creation has been filled with violence, disease, and death. People can believe in that kind of Little Tinkerer if they want to, but for honesty's sake they really should not call it a Christian view of GOD.

It takes a lot of arrogance to believe that our twenty-first century materialistic understanding has to be right or GOD is a deceiver. GOD has told us the truth, but many choose to believe their own very narrow materialistic understanding instead of what GOD says. They would be wise to wait until they had comprehensive knowledge of our universe and of life before jumping to premature materialistic conclusions regarding origins. Arrogance towards and ignorance of the truth have often been humanity's sinful problem; a problem that can be overcome only through humility, confession, and faith in CHRIST and His Word.

If their accusation were true, GOD would be prohibited from doing anything that is truly miraculous because scientific materialism limits GOD to working only through what is natural according to our contemporary state of knowledge. They are being very narrow minded about what GOD can and cannot do. The historical testimony that GOD has given us in His Word on creation, the flood, the crossing of the Red Sea, the raising of the dead, the virgin birth, and the resurrection of CHRIST is true regardless of the present state of scientific materialistic understanding. GOD is in no way limited to our very narrow comprehension of the physical realities of our universe. GOD'S science is infinitely higher than ours.

GOD IS THE MIRACLE WORKING GOD

The miracles recounted in the Scriptures are not deceptions of GOD but revelations of His infinite wisdom, power, and love; this includes the initial colossal miracle of creation itself.

GOD, FATHER, SON & HOLY SPIRIT, is a miracle working GOD. He always has been and always will be. GOD does for us what we cannot do for ourselves. If people want to limit GOD to present day scientific materialism, then it is not just the miracles of creation and the flood which they have to reject, but absolutely every miracle in the Bible, including the incarnation of CHRIST and His resurrection from the dead. Present day scientific materialism teaches us that none of these things can happen in our world. This is precisely the position of many of the liberal theologians over the past couple of centuries. Most of them have logically rejected all the miracles of the Scriptures and ended up believing in some kind of Gnostic or deistic version of GOD.

Take any miracle that GOD describes in His Word: is it not a deception according to "scientific materialism"? When JESUS turned water into wine in an instant, He did what normally takes months to do, growing the grapes followed by the process of fermentation. Was He being deceitful when He did it instantaneously? When He calmed the storm at sea, He did in a moment what normally follows over a period of minutes or even hours. Was He being deceitful? When He raised Lazarus from the dead, He did what medical science even today cannot do. Was He deceiving people into believing He could raise the dead?

It isn't logical to reject the teaching regarding the creation of our world, and to believe the other miracles that GOD'S SON and SPIRIT and faithful church have testified to in His Word. If one rejects the very clear revelation on creation because of present day scientific understanding, to be consistent an individual must also reject all other miraculous acts of GOD that are beyond contemporary science. All the miraculous events testified to in the Bible require a unique act of GOD. As JESUS said: ***"With man this is impossible, but with GOD all things are possible." (Matthew 19:26)***

The creation of our world is no more contrary to science than is the first coming of CHRIST and His promised return. We believe, not because we have a scientific explanation of how GOD has accomplished or will accomplish these miraculous events, but because through CHRIST and the HOLY

SPIRIT we know that GOD is the GOD of truth. We were not around six thousand years ago when GOD created our universe and world and we may not be living in this world when CHRIST returns. Both teachings are accepted on the basis of faith in the trustworthiness of GOD speaking through His SON, SPIRIT and Word. This is what faith is.

THE EVOLUTIONARY SCENARIO HAS NO ROOM FOR CHRIST IN ITS INN

There is a lot of propaganda put out by the evolutionary and, particularly, theistic evolutionary voices that tries to sell the idea that what a person believes about evolution or creation has no relationship to their belief in JESUS CHRIST. They are playing on the gullibility of people and the desire for peace and harmony in society above truth in the inner being. It is crucial for all of us that we are honest with ourselves and with GOD.

As was stated in chapter four it is true that just because people may believe in evolution does not mean that they cannot believe in JESUS. Likewise, just because someone believes in the Genesis account of creation does not necessarily make them a disciple of CHRIST. Consistency and faithfulness to GOD is not humanity's forte. Only GOD can judge an individual's heart. No one has perfect faith, and none of us is completely consistent and faithful in following CHRIST or believing and obeying His Word. We are not saved because our faith is perfect or because we have a perfect theological understanding of all parts of GOD'S Word. If our faith had to be perfect, then no one would be saved. Rather, we are saved because of our faith in a truthful and gracious GOD revealed through the atoning work of CHRIST and through the HOLY SPIRIT'S testimony in His Word.

Even though we are not saved through faith in the Genesis account of creation, it is a complete cover up to say that belief in evolution does not have any impact on people's faith in the reality of CHRIST and His Gospel. The upfront truth is that the evolutionary scenario has no room or cradle for JESUS in its inn. The real JESUS in the Gospels, who did miracles, spoke about heaven and sin and the judgment of hell, just does not fit in with the materialistic foundation of the evolutionary Tower of Babel. There is no place for JESUS, the SON of GOD and the descendant of Adam and Eve in their evolutionary mythology. JESUS would be nothing more than an evolved ape-man. There would be no historical SON of GOD, no prophecies of a coming Messiah, no virgin birth, no miracles, no cross, no resurrection, and therefore, no Gospel. If you remove the miraculous from JESUS' life, there would be no JESUS left to save anyone.

WHERE THERE IS NO SIN THERE IS NO NEED FOR A SAVIOR

One of the core beliefs of evolution is that humanity has evolved out of the ape kind. In the evolutionary mythology there was no special creation of Adam and Eve, no garden of Eden, no fall into sin, no death as a result of sin. As was already shown in chapter five, death is seen as part and parcel of the evolutionary process. Violence, abuse, injustice, lying, stealing, and killing would all have been part of the legacy of the ape-man kind and the animal kingdom out of which they believe that humanity emerged. If there was no fall from the glory of GOD, the image in which GOD created us, there would be no need of a Savior to save us. We would not need a Savior to save us from something that never happened. A prominent atheist, Frank Zindler, said in a debate with Christian apologist Dr. William Craig:

> *The most devastating thing though that biology did to Christianity was the discovery of biological evolu-*

tion. Now that we know that Adam and Eve never were real people the central myth of Christianity is destroyed. If there never was an Adam and Eve there never was an original sin. If there never was an original sin there is no need of salvation. If there is no need of salvation there is no need of a saviour. And I submit that puts JESUS, historical or otherwise, into the ranks of the unemployed. I think that evolution is absolutely the death knell of Christianity.[77]

Why is it that unbelievers can often clearly see what many who call themselves Christian are blind to? Many do not want to see the truth that the evolutionary mythology would completely eliminate the Gospel of CHRIST, because that would mean that they would have to choose between CHRIST and the evolutionary beliefs. Choices require convictions, convictions require understanding, and understanding requires seeking after the truth. It is much easier to pretend that evolution and CHRIST can somehow fit together than to confront the obvious conflict. It is easier to go with the flow of the world than to swim against the current.

At the heart of the Gospel of CHRIST is the message that GOD in His gracious love gave us His SON to be our Savior from sin, death, and hell by dying on the cross for our sin and rising from the dead, to give all who trust in Him the gift of eternal life in His coming Kingdom. The evolutionary mythology of humanity's origin rips out the heart of the Gospel message by denying the fact of sin, and therefore death as the result of human sin. Since the central work of CHRIST in His death on the cross is to forgive sin His work would be completely void of purpose if humanity's fall into sin were not true.

It is written:

"Therefore, just as sin entered the world through one man, and death through sin, and in this way death came to all men, because all sinned—for before the law was given, sin was in the world. But sin is not taken into account when there is no law. Nevertheless, death reigned from the time of Adam to the time of Moses, even over those who did not sin by breaking a command, as did Adam, who was a pattern of the one to come. But the gift is not like the trespass. For if the many died by the trespass of the one man, how much more did God's grace and the gift that came by the grace of the one man, Jesus Christ, overflow to the many! Again, the gift of God is not like the result of the one man's sin: The judgment followed one sin and brought condemnation, but the gift followed many trespasses and brought justification. For if, by the trespass of the one man, death reigned through that one man, how much more will those who receive God's abundant provision of grace and of the gift of righteousness reign in life through the one man, Jesus Christ." Romans 5:12-17

In these words GOD'S HOLY SPIRIT explains the good news of JESUS CHRIST in direct contrast to the sin and death that entered the world as a direct result of Adam's disobedience to the command of GOD. CHRIST'S death is meaningless if Adam did not exist or if Adam's sin did not result in ushering sin and death into our world. There is no need of grace where there is no sin. Consequently, under the evolutionary scenario the whole of the Gospel is nothing but an empty shell.

With no Gospel message of CHRIST to proclaim, theistic evolutionists often occupy themselves, or soothe themselves and others, by trying to create some form of spirituality and religious humanistic ethic from the ruins in which evolution has left their faith. Worshipping the goddess Sophia, replacing the cross with more positive symbols, creating labyrinths for meditation, enduring the sweat lodge, consulting imams and yogis, following yoga rituals, rearranging their homes in conformity with the spirituality of feng shui, reading the crystals, and other mystic new age practices are all part of the search to find something, anything, to fill in the empty void created by their rejection of the historical CHRIST. Through faith in the mythology of evolution many have lost their faith in CHRIST and His Word. They should resign their positions in Christian churches and begin their own religion of "Little Tinkerism" or "evolutionary humanism".

JESUS said: "Watch out for the false prophets. They come to you in sheep's clothing" (Matthew 7:15)

THE TRUTH IS MARCHING ON AND SETTING PEOPLE FREE

If GOD is not the GOD of truth, then the evolutionary narrative might stand a chance. GOD is truth, and therefore evolution isn't.

CHRIST, the SPIRIT and the Word do reveal the reality of a being who is a liar and a murderer. JESUS testified: *"He was a murderer from the beginning, not holding to the truth, for there was no truth in him. When he lies he speaks his native language, for he is a liar and the father of lies." (John 8:44)* Through the evolutionary mythology, satan has spun a web of lies that lead people and nations down the path of destruction. The fruits of evolution fulfill these words of CHRIST, as we will highlight in the next chapter.

In John 8, JESUS said: **"If you hold to my teaching, you are really My disciples. Then you will know the truth, and**

the truth will set you free". **(John 8:31,32)** We praise GOD that through the Word of CHRIST He is setting people free from all of the mythologies of our world, past and present. It is crucial for those who believe in CHRIST to hold to His teachings, including His teachings on creation and the flood. Only by standing firm in His Word can we be set free from the evolutionary mythology."The Battle Hymn of the Republic" written by Julia Ward Howe, proclaims:

Glory, glory, hallelujah!
Glory, glory, hallelujah!
Glory, glory, hallelujah!
His truth is marching on.

PART C
FRUITS AND CHOICES

CHAPTER 7

By Their Fruits

JESUS' WORDS OF WARNING

JESUS said: "Watch out for false prophets. They come to you in sheep's clothing, but inwardly they are ferocious wolves. By their fruit you will recognize them. Do people pick grapes from thorn bushes, or figs from thistles? Likewise every good tree bears good fruit, but a bad tree bears bad fruit. A good tree cannot bear bad fruit, and a bad tree cannot bear good fruit. Every tree that does not bear good fruit is cut down and thrown into the fire. Thus, by their fruit you will recognize them." Matthew 7:15-20

If these words had been spoken by anyone else we would not have to pay attention to them, but they were spoken by JESUS, the CHRIST, who Was, Is, and always will be GOD'S SON and humanity's one and only SAVIOR and LORD. To those who believe in JESUS His words are crucial and need to be comprehended and applied. A few verses later, JESUS taught that those who hear His Word and practice it in their lives are wise. A foolish person, on the other hand, hears

CHRIST'S Word but does not practice it. The final result is that the life of the wise is secure despite the storms or troubles encountered, whereas the life of the foolish crumbles when the tempests rage. Life certainly has its share of tumult for all of us. Therefore it is wise to pay careful attention and practice GOD'S Word, just as JESUS said.

CHRIST'S words here are very sombre words that speak of great deceivers in our world, specifically of those who come camouflaged in Christian faith. This is not the only time when GOD warns us of the dangers of false teachers and teachings. The LORD does this repeatedly from the beginning to the end of His Word- from Genesis to Revelation. Like parents warning their children of the dangers in life, GOD warns His people of the perils our faith will face in this world.

Particularly relevant for us is CHRIST'S reiteration and emphasis on false teachings in His prophecy on His second coming. Three times in Matthew 24:1-25, where JESUS speaks about the events and signs preceding His return to our world, He commands us to watch out for the false CHRISTS, false prophets, and false teachings that will deceive and draw many away from faith in Him. Do not underestimate these false teachers and leaders. They will be very impressive and convincing to those who do not continue in GOD'S Word. Anyone who is willing to exchange GOD'S truth for the world's wisdom, in order to be approved by the politically correct crowd, will be deceived by one or more of these false teachings and leaders. Only GOD knows the precise date of JESUS' return, but certainly the signs around us should cause us to be vigilant, testing everything by GOD'S truth and GOD'S love.

THE FRUITS OF THE EVOLUTIONARY TREE

JESUS said that we could identify false teachings and teachers by their fruits. What are the fruits of the teaching of evolution in the church and in the world around us?

Biological evolution is often portrayed as a tree with one root, and one stem with many branches. This picture of Darwin's evolutionary tree communicates the belief that all life has come from a single root, a single cell, and has merely grown and branched out from there into the great diversity of life on our planet.

The evolutionary tree, which has spread its roots and branches to every corner of the earth, produces a lot of poisonous fruit. Evolution is a wild tree that flourishes on the fertilizer of the belief in "billions of years of suffering and death". Its fruit corresponds with its fertilizer. As with computers, the acronym GIGO applies - "garbage in – garbage out." The evolutionary tree produces suffering and death in any nation where it takes root and spreads its influence. The fruit produced by the evolutionary tree comes in many colours, shapes, sizes and tastes: unbelief in GOD and His Word, atheism, idolatry, abortion, euthanasia, racism, eugenics, foetal experimentation, death camps, sexual promiscuity, including homosexuality, and the destruction of the institution of the family, to list a few. To be sure, there are many other trees in our world that also produce these same destructive fruits, and cross pollination happens frequently. Evolution is not the only wild and toxic tree around; it is just one that has grown incredibly huge in our contemporary world.

CONTEMPT FOR GOD, HIS WORD AND HIS CHURCH

Many people who live in the shadow of the evolutionary tree turn from belief in the GOD of the Bible to a rejection and growing contempt for GOD, His Word and His Church. When we refer to GOD as the GOD of the Bible, we are highlighting the relationship between GOD and His Word. GOD is the ultimate author of His Word. JESUS prayed for His Church: *"Sanctify them in Your truth, Your Word is Truth." (John 17:17)*

Not surprisingly, wherever the evolutionary tree spreads its branches there is a corresponding decrease in trust and love for the Holy GOD, His Word and His Church. The shadows from the branches of the evolutionary tree create a dark place where the light and warmth of GOD are not seen or felt. Where life is seen as the accidental result of billions of years of chance processes, producing the ravages of suffering and death, there is no comprehension of a Holy GOD of love and truth. Rather than GOD being sought after and loved, He is seen as contemptible and irrelevant. Consequently, the Bible is viewed as nothing more than an ancient book of myths that deceives and divides people.

Tragically, this is also true in many churches that have taken up residence beneath the evolutionary tree. Once church leaders and individuals succumb to the evolutionary myth of origins, they foolishly think that they are released from adhering to the Bible as GOD'S Word, and set free to develop and spread their own word and version of GOD to the world. The world is littered with the skeletons of denominations and churches which once believed in the Holy GOD revealed in the Scriptures, before swallowing the poison of evolution and turning to a god of their own imaginations.

In 1997 Rev. Bill Phipps, the moderator of the largest protestant church in Canada, the United Church of Canada, openly declared in an interview with the Ottawa Citizen: "I don't believe Jesus was God...I don't believe Jesus is the only way to God...I don't believe he rose from the dead as a scientific fact. I don't know whether those things happened. It's an irrelevant question."[78] He also stated that he didn't know if there was a heaven or a hell.

How is it possible that the leader of a Christian Church, whose central belief about GOD was that He has revealed Himself to our world through His SON JESUS CHRIST who came to save and to give eternal life, openly professes his unbelief that CHRIST was and is GOD? Not only did he acknowledge his rejection of JESUS, but the national board of the United Church supported his position as falling within the wide range of views accepted and celebrated by the denomination. How did this happen? It didn't happen overnight. For the past seventy years an increasing number of United Church leaders and clergy have ceased to believe in the Holy GOD of the Christian faith and have come to believe in the Little Tinkerer goddess of evolution. This belief is like gangrene that spreads and destroys the very faith that they began with, and that has been proclaimed for two thousand years.

What is true of the United Church of Canada has become a visible feature in many, if not all, of the liberal denominations throughout the world. Denominations that absorb the evolutionary mythology into their worldview begin with the rejection of the first few chapters of Genesis, the creation and the flood, and from there the dominoes of Biblical teachings fall one after another, until virtually everything of the Christian faith and doctrine is gone, including the divinity of CHRIST. In the Evangelical Lutheran Church of Canada seminary I attended, the liberal indoctrination began with the rejection of the creation account in Genesis, and proceeded from there to doubting the historicity of much of Scripture including the virgin birth, miracles and even the words of CHRIST, all because the "enlightenment" of the scientific approach to life demanded it.

The HOLY SPIRIT inspired testimony to the resurrection of CHRIST is just as unscientific as the Genesis account of the creation of the universe. Once the miracle of creation is denied as a result of belief in evolution, the "demythologizing" of all of the miracles of Scripture follows. No wonder the pews of theistic evolutionary churches are being deserted. JESUS warned His followers about many of the religious leaders of His day: *"He replied, 'Every plant that My Heavenly Father has not planted will be pulled up by the roots. Leave them; they are blind guides. If a blind man leads a blind man, both will fall into a pit.'" (Matthew 15:13,14)*

The rejection of the historicity of Scripture, as a result of the acceptance of evolutionary mythology, is naturally accompanied by a rejection of the theological, moral and ethical teachings GOD gives us in His Word. The sanctity of human life is being erased by the evolutionary shadow. If the unborn are not recognized as created in the image of the Holy GOD, but are viewed as nothing more than animal foetuses or little pieces of evolved flesh, then their lives are worthless; even worse, they are seen as burdens that can be eliminated. It is not earthshaking that the vast majority of liberal denominations which have imbibed evolution have been in the forefront of efforts to legalize and even encourage the abortion of unborn children. In Canada it is doubtful if the law legalizing abortion would have been passed in 1969 without the public support of the two largest liberal denominations, the United Church of Canada and the Anglican Church of Canada. In the United States many of the liberal denominations have their own hospitals that "minister" to women by "relieving them of the burden" of their unborn children. Rather than caring for these orphans (James 1:27) or encouraging their adoption into loving homes, they tear them out of their mothers' wombs and discard them with the garbage. Churches and hospitals which were established to save lives have been transformed into death chambers through the poison of evolution.

Liberal denominations today are just following the crowd of evolutionary-minded liberal thinkers in the acceptance and promotion of all forms of sexual immorality, including homosexuality. Once human beings are seen through the evolutionary prism as evolved animals, they imagine that they are set "free" from the outdated and oppressive sexual morality of the Christian faith to experiment with all kinds of sexual desires and fantasies. Under the influence of the evolutionary poison, the institution of marriage given by GOD in Genesis is seen as nothing more than an ancient patriarchal view out of which we have now evolved and been set free. Homosexual marriage is seen as the latest step in our evolutionary development. GOD said through the prophet Isaiah: **"Woe to those who call evil good and good evil, who put darkness for light and light for darkness, who put bitter for sweet and sweet for bitter." (Isaiah 5:20)**

Evangelical denominations that are now embracing the evolutionary mythology in their seminaries and Bible schools believe they are immune to the drift away from CHRIST and His Word that has rotted many of the mainline churches. Unless they turn back, within a few years their positions will align with that of the liberal denominations today. The evolutionary venom of unbelief will spread all the way to the identity of CHRIST, while they argue over budgets and whether global warming is real or not.

It is written:

"For the time will come when men will not put up with sound doctrine. Instead, to suit their own desires, they will gather around them a great number of teachers to say what their itching ears

want to hear. They will turn their ears away from the truth and turn aside to myths." (2 Timothy 4:3,4)

There is no greater example of these verses today than the rejection of the Biblical revelation on creation to embrace the evolutionary myth. JESUS said to those who believe in Him: ***"You are the salt of the earth. But if the salt loses its saltiness, how can it be made salty again? It is no longer good for anything, except to be thrown out and trampled by men." (Matthew 5:13)*** The Christian church is called to be the salt of the earth but once it loses its saltiness it is of no use to GOD'S eternal Kingdom and will be thrown out and trampled underfoot by the world. Apart from a great revival, a repentance and return to CHRIST and His Word, the fate of all of the mainline churches is already sealed and many "evangelical" denominations will be the next to succumb. Evolution is doing its deadly work in depleting their faith, and emptying their pews. One of the reasons so many people have rejected CHRIST and His church since Darwin's day is that they have chosen to believe in evolution. There is no sense in worshipping a god who would create using such a haphazard and cruel process. Little Tinkerer is not worth worshipping.

If this is what the poison of evolution does within the walls of the church, what is it doing in the world around us?

ATHEISM: THE LOGICAL FRUIT OF EVOLUTION

It is written: "The fool says in his heart: 'there is no god.'" Psalm 14:1

Although history records the presence of individuals and even certain philosophical and spiritual groups who held to some forms of atheism, it was never widely embraced in any of the pre-nineteenth century civilizations. However, with the advent of evolutionary belief following Darwin, evolutionism and its fruit of atheism began to spread through the self-proclaimed intellectual elite to all corners of our world, beginning in Europe. Although the evolutionary tree is not the only tree that bears atheism, it certainly has been the most fertile tree. Before we turn to evolution's seduction of nations into pagan attitudes, values, and deadly practices it is vital to understand the logical impact of evolution on many individuals.

Charles Darwin grew up in England, attended an Anglican school and was initially preparing to become an Anglican priest. However, he chose to move away from his cultural Christian upbringing and to reject the historical witness of the Bible. Finally, as a result of his growing fascination with the development of the theory of evolution, and his recognition of all of the cruelty and suffering of life on our planet, including the death of one of his daughters, he completely rejected the reality of an omnipotent loving GOD. By the time of his death he was either a deist (someone who hypothesizes an irrelevant absentee impersonal god with no concern for the world), an agnostic (someone who claims to have no opinion concerning god), or an outright atheist, depending on which scholar's interpretation is most accurate. There has been some speculation that he had a "deathbed repentance", but there is little solid historical information to substantiate that claim. GOD knows.

Like Darwin many, probably most, of those who are entranced by evolutionary mythology end up as deists, agnostics, atheists, or, over the last century, have turned to some mystical or new age view of god as a force or energy cycling through the universe and in and through them. This is especially true for many scientists who have come under the influence of evolutionary biology. Thankfully, there are tens of thousands of individuals in scientific fields or scientific related fields such as medicine, engineering, and computer science who confess a Christian Biblical faith. Nevertheless, the statistics certainly seem to indicate that the scientific community has a much higher percentage of atheists, agnostics, and deists than the general population.

> The scientific community, above any other subgroup of the population, has become overwhelmingly atheistic. According to a <u>1998 report in Nature</u>, a recent survey finds that, "among the top natural scientists, disbelief is greater than ever; almost total". Interestingly, the biologists in the National Academy of Science possess the lowest rate of belief of all the science disciplines, with only 5.5% believing in God. This decline in belief in biologists strongly indicates the nature of the cause, and the ability of the teaching of evolutionary biology to turn people away from a belief in God.[79]
> Also in a <u>1991 Gallop poll a clear trend emerged</u> demonstrating that higher education, and belief in evolution over millions of years as the source of human existence were concurrent. From these statistics it would appear that higher education, and particularly specialization in the natural sciences will indoctrinate students into a naturalistic or atheistic view of the world...Education in these naturalistic philosophies, and the pervasive teaching of evolution is almost certainly the principal influence affecting the rise of atheism in our scientific community. Evolution may be better called <u>evolutionism as it is considered a religion</u> by many. Evolution is the champion theory of secular humanism, and a scientific community now totally under the control of an atheistic majority. Evolution is being used in attempt to explain the origin and evolution of life on earth

without a supernatural creation. These theories are being taught as a matter of fact in science classes today, and such teaching will affect the way people view the world. If they are left unchallenged, this inundation will cause belief in God as the source of life to diminish, and evolution ultimately has the power to convince people there is no God.[80]

Dr. Jonathon Sarfati, in an article called " National Academy of Science is godless to the core — *Nature* Survey" on the Creation International website, wrote:

But a recent survey published in the leading science journal Nature conclusively showed that the National Academy of Science is anti-God to the core. A survey of all 517 NAS members in biological and physical sciences resulted in just over half responding. 72.2 % were overtly atheistic, 20.8 % agnostic, and only 7.0 % believed in a personal God. Belief in God and immortality was lowest among biologists. It is likely that those who didn't respond were unbelievers as well, so the study probably underestimates the level of anti-God belief in the NAS. The unbelief is far higher than the percentage among scientists in general, or in the whole population.[81]

Although many polls are easily manipulated to say whatever the interpreter wants them to say, these polls are certainly consistent with the personal testimony of many individual scientists, particularly evolutionary biologists. Dr. Will Provine, professor of biology at Cornell University, wrote:

Evolution is the greatest engine of atheism ever invented[82]

My observation is that the great majority of modern evolutionary biologists now are atheists or something very close to that. Yet prominent atheistic or agnostic scientists publicly deny that there is any conflict between science and religion. Rather than simple intellectual dishonesty, this position is pragmatic. In the United States, elected members of Congress all proclaim to be religious; many scientists believe that funding for science might suffer if the atheistic implications of modern science were widely understood. [83]

Let me summarize my views on what modern evolutionary biology tells us loud and clear — and these are basically Darwin's views. There are no gods, no purposes, and no goal-directed forces of any kind. There is no life after death. When I die, I am absolutely certain that I am going to be dead. That's the end of me. There is no ultimate foundation for ethics, no ultimate meaning in life, and no free will for humans, either. [84]

Bertrand Russel, one of the most popular atheist writers and philosophers of the 20th Century, wrote:

Such, in outline, but even more purposeless, more void of meaning, is the world which Science presents for our belief. Amid such a world, if anywhere, our ideals henceforward must find a home. That Man is the product of causes which had no prevision of the end they were achieving; his origin, his growth, his hopes and fears, his loves and his beliefs, are but the outcome of accidental collocations of atoms; that no fire, no heroism, no intensity of thought and feeling, can preserve an individual life beyond the grave; that all the labours of the ages, all the devotion, all the inspiration, all the noonday brightness of human genius, are destined to extinction in the vast death of the solar system, and that the whole temple of Man's achievement must inevitably be buried beneath the debris of a universe in ruins.[85]

Dr. Richard Dawkins, an evolutionary biologist and formerly Professor for Public Understanding of Science at Oxford, has become a rabid anti-Christian and anti-GOD author who has been actively and successfully promoting atheism in the western world for over a quarter of a century. Although he was raised with a nominal Christian background, by his mid-teens, after being taught the doctrines of evolution, he came to believe that evolution was the only scientific understanding of life, and that atheism was the only logical conclusion of evolution. In his 1986 book *"The Blind Watchmaker"* Dawkins wrote: "although atheism might have been logically tenable before Darwin, Darwin made it possible to be an intellectually fulfilled atheist."[86]

Why is atheism a logical conclusion of evolution? Dawkins puts it this way:

The total amount of suffering per year in the natural world is beyond all decent contemplation. During the minute that it takes me to compose this sentence, thousands of animals are being eaten alive, many others are running for their lives, whimpering with fear, others are slowly being devoured from within by rasping parasites, thousands of all kinds are dying of starvation, thirst, and disease. It must be so. If there ever is a time of plenty, this very fact will automatically lead to an increase in the population until the natural state of starvation and misery is restored.

> *In a universe of electrons and selfish genes, blind physical forces and genetic replication, some people are going to get hurt, other people are going to get lucky, and you won't find any rhyme or reason in it, nor any justice. The universe that we observe has precisely the properties we should expect if there is, at bottom, no design, no purpose, no evil, no good, nothing but pitiless indifference.[87]*

Charles Templeton was a well-known Christian evangelist, a pastor of a growing evangelical church in Toronto, one of the first vice-presidents of Youth for CHRIST ministry, and a friend of world evangelist Billy Graham. He testified to the Gospel of CHRIST before tens of thousands of people. Nevertheless, as he came to faith in evolution as the explanation for the origin and development of life, he gave up his faith in CHRIST and in the truth of GOD'S Word. He did what Esau did, and exchanged his birthright for a mess of pottage.

GOD warns us: **"So, if you think you are standing firm, be careful that you don't fall!"(1 Corinthians 10:12)** Even the mighty, if they don't stand firm on GOD'S Word of truth, can fall before the powerful myth of evolution which comes disguised in the mantle of science. Templeton, like Darwin and millions before and after him, gave up his faith in the Biblical teaching of creation and humanity's plunge into sin, and was left with a cruel cold dying world in which there were no comforting arms of a loving GOD, and no hope of an eternal paradise to come. Without GOD and CHRIST and the teachings and promises of His Word there is no hope for our lost race.

If anyone doubts the power of evolution to fertilize the soil that creates atheists, all he or she has to do is peruse the blogs of most atheist and evolutionist websites. There a person can read testimony after testimony of those who state that they once believed in GOD but because of their trust in evolution have given up their faith. Of course, a careful reading of their testimonies reveals that they often have only an elementary or shallow understanding of evolution and how it works, and virtually no real understanding of GOD and His Word. As we said in chapter three, most people who believe in evolution do so, not because they really comprehend it but because that's what the "scientific experts say". Though they really do not understand it, they feel justified in using evolution as a rationalization for living a godless life. They have chosen to trust in their own understanding and the wisdom of people instead of trusting what GOD has said in His Word. This becomes a dominant motif in their life, a fatal obsession, and a force of unbelief that propels them down the broad way that leads to death.

Here is an opposite testimony by Gil Dodgen, a fanatical atheist at one time who came to know CHRIST in a living and personal way:

> At risk of being labeled a religious fanatic who has abandoned all powers of reason, I present the following, written in 1994 at the behest my friend David Pounds, who is mentioned. It was David who recommended that I read Michael Denton's *Evolution: A Theory in Crisis*.
>
> In answer to the question, Does Atheism Poison Everything? — I can attest to the fact that it does. It poisoned my life, soul, and intellect for 43 years.
>
> Readers can take the following as they will. I present it as my experience, which transformed my life, intellect, and soul much for the better. My former atheism was universally and consistently destructive in all three categories.

A Christian Testimony
March 12, 1994 at age 43
Gil Dodgen

All my life I was an atheist, and believed that I was just a complex piece of biochemistry that came about by chance. When my chemical processes shut down at death I would cease to exist and enter eternal oblivion. I had (and have) everything the world has to offer — a wonderful wife, two beautiful children, a nice home and a job I enjoy — but somehow, as I grew older, I increasingly sensed that something was missing. I started asking myself, "What is all this for in the end? Is there some purpose behind all this? Are my children nothing more than a bunch of chemistry as well?" Deep inside I knew there had to be more.

One day, while shopping in a book store, I ran across a cartoon video entitled "The Lion, The Witch And The Wardrobe." The blurb on the back of the package made it look like a nice story for my five-year-old daughter, and it was only $9.95, so I bought it. One evening my daughter and I lounged on the living room floor and watched the cartoon. At one point the lion gave up his life to an evil witch to spare a child, on whom the witch had a claim. The story of his courage, compassion and unjust death moved me deeply, and his return to life and struggle against evil inspired me.

Suddenly I realized that the lion was Jesus Christ. There had been no mention of the allegory on the package, but I knew enough about the life of Jesus from my secular studies to recognize it. In retrospect I realize that Someone had gone to work on my heart as well.

I had a friend, Dave Pounds, and I knew that he was a devout Christian. (In fact, he was the only Christian I knew.) We had not seen each other in a long time and I thought the video might make a nice excuse for our families to get together for dinner, so I gave his wife a call. The evening came and we all watched the video. When it came to the death of the lion I started asking Dave questions about how certain parts of the cartoon related to the story of Jesus, and decided that it might be interesting to read about it in the Bible. He went out to his car and gave me one.

For the next few weeks I read the New Testament, and began to feel a transformation taking place. I had no idea how much wisdom there was in this book, and how beautiful and inspiring the life of Christ was. I was particularly moved by the story of His death, and His total devotion to God, righteousness and love of His fellow man. How could He forgive His murderers from the cross? His final words were burned into my mind: "Father, into your hands I commit my spirit."

A battle had begun — an intense mental conflict that lasted a few weeks. I would argue with myself, "This is just a bunch of mythology; you're too smart to believe this. No, this is good, this is true, this is beautiful. There is a God, and He loves you. Jesus died a hideous death for you on the cross." I noticed, however, that once I started reading the Bible the arguing ceased.

I would call up Dave to ask questions about what I had read in the Bible, and one evening he prayed for me over the phone, and suggested that I try it myself. Although prayer was a totally foreign concept to me, I felt that somehow I had to give it a try. So, with the greatest sincerity I could muster I knelt in prayer. In Jesus' name I asked God to reveal Himself to me, to bring Christ into my life, to lead me where He wanted me to go. I knew at this point that there was no going back; I just couldn't return to the darkness of my former atheism. But I still lacked the courage of total commitment. I was still in conflict.

Then one night I went to a concert at Calvary Chapel, Costa Mesa, California. The master of ceremonies was Warren Duffy, a Christian radio talk show host from whom I had heard about the concert. He asked everyone to do something that they used to do in his church when he was a pastor. He asked us to turn to someone we didn't know, give them a big hug and say, "God loves you and I do too." I turned to the guy standing to my right, and hugged him. He said, "God loves you and I do too." I hesitated just a moment and then said it back to him. The strange thing was, I meant it.

At that moment I was overwhelmed with joy and a complete sense of peace, as I felt what I knew was Christ's perfect love overtaking me. It was as though a pair of spiritual eyes, that I didn't even know I had, were opened. Suddenly I understood — in an indescribable and profoundly spiritual way — what Christ had done on the cross. As He hung there He was putting His arms around a stranger, saying, "I love you Gil, and my Father who sent me does too." He meant it, and proved it by suffering and dying for me.

There was no mistaking it, I had become a Christian, and in silent prayer I accepted Jesus Christ as my Lord and Savior. The battle was over.

I have left the dark, cold, depressing, nihilistic depths of atheism, and have been brought into the light, warmth, joy, peace and love of Jesus Christ. The missing piece in my life has been found. I think about my Savior almost always, whether consciously or in the back of my mind. I know that He is with me, even when it might not seem like it. I have new, Christian standards for my life, and although I haven't lived up to all of them, I have Someone on my side to help me work toward these goals.[88]

NO GODS, MANY GODS, OR THE ONE AND ONLY GOD?

One of the mantras frequently recited by those who are atheistic is that they simply believe in one less god than Christians do. In order to maintain a materialistic evolutionary worldview, atheists have to reject all beliefs in all the gods or goddesses of the various religions. Christians recognize that the world is filled with many man-made gods; atheists believe that GOD is just a human fabrication like all of the other gods. What they are unwilling to comprehend is that the human creation of many false gods or idols does not imply that there is not a true living GOD, the one and only Creator of the heavens and the earth.

It is written:

"Not to us, LORD, not to us, but to Your name be the glory, because of Your love and faithfulness. Why do the nations say, "Where is their GOD?" Our GOD is in Heaven; He does whatever pleases Him. But their idols are silver and gold, made by human hands. They have mouths, but cannot speak, eyes, but cannot see. They have ears, but cannot hear, noses, but cannot smell. They have hands, but cannot feel, feet, but cannot walk, nor can they utter a sound with their throats. Those who make

them will be like them, and so will all who trust in them." (Psalm 115:1-8)

It is wise to reject all false gods, but it is foolish to reject the one true living GOD. Further, it is irrational to think that because there are many falsehoods that there must not be any truth. If that were true, science would be impossible. There are always an unlimited number of falsehoods for every truth. Falsehoods do not eliminate truth, but highlight its importance.

In contradistinction to atheists, there are many who through uncritical thinking have come to the conclusion that it is possible to believe in all of the gods of the various religions at the same time. This polytheistic position has been adopted by many of the theistic evolutionists, who do not have a clear understanding of the nature of GOD, and therefore find it easy to mix and match GOD to fit almost all beliefs. The true character of GOD is not important to them; all that matters is that there is peace between religions and peace between religion and science. They are willing to sacrifice the nature of GOD for religious peace on earth.

When individuals, whether atheists or theists, use the argument that all religions are equal, the subtext hidden underneath is that they are all equally wrong, since there is absolutely no way they could all be right. Most religions are polytheistic, believing in the existence of many gods, but Christianity, Judaism, and Islam are monotheistic, proclaiming that there is only one GOD. Both cannot be true. The heart of our Christian faith is the reality that JESUS CHRIST was and is the eternal SON of GOD. However, both Judaism, for the past two thousand years, and Islam, for the past fourteen hundred years, have clearly rejected JESUS as the eternal SON of GOD. Any rational reading of the Gospels indicates that JESUS was crucified precisely because He claimed to be the eternal SON of GOD, which to the Pharisees and Sadducees was absolute blasphemy, just as it is for orthodox Jews and Muslims today. It is dishonest to not acknowledge the fundamental differences between our Christian beliefs and the other beliefs present in our world. Although there are millions of false idols, there is only the one Living GOD and His SON JESUS CHRIST, humanity's only Savior. We are to love our neighbours whether they are atheists, Hindus, Moslems, or new agers, but part of loving them is sharing the grace and the truth of CHRIST with them. The prophet Elijah, speaking for GOD to people who had similar fuzzy thinking two thousand eight hundred years ago, said: *"How long will you waver between two opinions? If the LORD is GOD, follow him; but if Baal is god, follow him." But the people said nothing." (1 Kings 18:20-22)*

THE BLOODY RECORD OF THE 20ᵀᴴ CENTURY

"Why do the nations conspire and the peoples plot in vain? The kings of the earth take their stand and the rulers gather together against the LORD and His Anointed One, saying, "Let us break their chains and throw off their fetters. The One enthroned in Heaven laughs; the LORD scoffs at them. Then He rebukes them in His anger and terrifies them in His wrath, saying, 'I have installed My King on Zion, my Holy Hill." Psalm 2:1-3

Although the logic of pure materialism, including evolution, is vacuous, without any foundation of ultimate origins, it continues to deceive individuals and entire nations. Whether it is in the form of communism, socialism, corporate capitalism, or secular humanism, evolution is the foundation on which materialists have built and are building their "utopian societies", all of which end in death, often mass exterminations. Western civilization is in the process of disintegration from within as a result of injecting this evolutionary toxin. Evolution is a necessity in our contemporary world for those who desire to live without the constraints on human activities that the Biblical historical Christian GOD brings.

Atheistic or agnostic individuals, particularly among the self-proclaimed intellectual elite, having given up any hope of eternity, often turn their sights on gaining power and influence for transforming their nations, by leading them into the godless abyss of evolution. Wherever evolutionism has worked its poison in nations, the result has been catastrophic. Nations that have propagandized their people with evolutionary mythology are the greatest purveyors of death in the history of our world. The record of the twentieth century has been deeply stained with the blood of the masses that has been spilled by and in nations that were seduced by evolutionism.

In his article "Death by Government" historian R.J Rummel lists an estimated number of people killed by various governments in the twentieth century. The top three states responsible for the deaths of approximately 120,000,000 people were the Russian Communists, the Chinese Communists, and Hitler's Nazis. All three of these nations were strongly influenced by evolutionism; two of them became outright atheistic as a result. Of course evolutionary mythology was not the only factor in leading these nations down the lethal path, but it provided the crucial philosophical foundation for their death march. The actual figures were:

61,911,000 Murdered: The Soviet Gulag State
35,236,000 Murdered: The Communist Chinese
20,946,000 Murdered: The Nazi Genocide State[89]

Other historians provide slightly different estimates, but what cannot be denied, no matter whose figures one uses, is that it has been the governments and nations that have been most seduced by evolutionism that have been the greatest destroyers of human life in the world's history. Let's take a closer look at one of them.

FROM DARWIN TO HITLER: CONNECTING THE DOTS

How was it possible that one of the most advanced and powerful nations in the nineteenth and first half of the twentieth century could have been so misled as to embrace the demonic policies of the Nazis? It didn't happen overnight. After the formulation of the evolutionary theory by Darwin in the middle of the nineteenth century, German intellectuals were amongst the first to imbibe its godless premises. As Darwinism spread, many of the German intelligentsia in the political, scientific, university, and even the theological communities began to embrace the evolutionary mythology. By the time Hitler arrived on the scene, many of the essential doctrines of evolutionism had been absorbed by many of the leaders, and propagandized to the people of the nation, through the educational and media institutions. In his book, *From Darwin to Hitler: Evolutionary Ethics, Eugenics, and Racism in Germany,* Dr. Richard Weikart, head of the department of history at California State University, Stanislaus, outlines the basic doctrines of Darwinism that had been baptized into the Weltanschauung (worldview) of Nazism. In an article called "Darwin and the Nazis" in the American Spectator, 2008 he wrote:

> As I show in meticulous detail in my book, the Nazis' devaluing of human life derived from Darwinian ideology (this does not mean that all Nazi

ideology came from Darwinism). There were six features of Darwinian theory that have contributed to the devaluing of human life (then and now):

1. Darwin argued that humans were not qualitatively different from animals. The leading Darwinist in Germany, Ernst Haeckel, attacked the "anthropocentric" view that humans are unique and special.

2. Darwin denied that humans had an immaterial soul. He and other Darwinists believed that all aspects of the human psyche, including reason, morality, aesthetics, and even religion, originated through completely natural processes.

3. Darwin and other Darwinists recognized that if morality was the product of mindless evolution, then there is no objective, fixed morality and thus no objective human rights. Darwin stated in his Autobiography that one "can have for his rule of life, as far as I can see, only to follow those impulses and instincts which are the strongest or which seem to him the best ones."

4. Since evolution requires variation, Darwin and other early Darwinists believed in human inequality. Haeckel emphasized inequality to such as extent that he even classified human races as twelve distinct species and claimed that the lowest humans were closer to primates than to the highest humans.

5. Darwin and most Darwinists believe that humans are locked in an ineluctable struggle for existence. Darwin claimed in The Descent of Man

that because of this struggle, "[a]t some future period, not very distant as measured by centuries, the civilised races of man will almost certainly exterminate and replace throughout the world the savage races."

6. Darwinism overturned the Judeo-Christian view of death as an enemy, construing it instead as a beneficial engine of progress. Darwin remarked in The Origin of Species, "Thus, from the war of nature, from famine and death, the most exalted object which we are capable of conceiving, namely, the production of the higher animals, directly follows."

These six ideas were promoted by many prominent Darwinian biologists and Darwinian-inspired social thinkers in the late 19th and early 20th centuries. All six were enthusiastically embraced by Hitler and many other leading Nazis. Hitler thought that killing "inferior" humans would bring about evolutionary progress. Most historians who specialize in the Nazi era recognize the Darwinian underpinnings of many aspects of Hitler's ideology.[90]

The rest is history. Eugenics in all of its ugliness mushroomed and smothered the vulnerable wherever they were found. Some contemporary evolutionists try to distance Darwin from the development of eugenics, the elimination of the weak and vulnerable for the good of the human race, but it was his closest supporters and family that began the movement. Darwin's cousin Francis Galton founded the eugenics movement in 1883 and Charles Darwin's son, Leonard, was president of the Eugenics Society in England from 1911 to 1928.[91] Eugenics, just one of the logical ends of all of this mythology, led to the slaughter of the handicapped, the mentally infirm, unborn children, gypsies, communists, Jews and anyone else that the Nazis deemed as useless eaters or accused of weakening the Arian race and standing in the way of their evolutionary march to superiority and power. Beliefs have consequences.

Where were the German church leaders during the time when all of this was developing around them? The vast majority of the German state churches had become so paralyzed by the liberal theology they had adopted that they had no strength to resist these evolutionary and Nazi doctrines. As a result many of the church leaders and members danced to the tune of the Nazis, who proceeded to efficiently eliminate all of the people that were labelled a hindrance to the nation and world. The institutional church in Germany has never recovered from their failure to be the Church of CHRIST. They have been trampled underfoot, just as JESUS said. To be sure there was a faithful remnant of believers who kept the faith and suffered through these terrible days in the history of their nation. Praise GOD for the faithful few.

All of the people said: "Never again!" However, as in the opening to "The LORD of the Rings" by J. R. R. Tolkien, that generation has now passed and the new generation has forgotten, history has become myth, and the poison of evolution continues to spread. There was about a quarter of a century interlude in Europe when anti-life and anti-Semitic attitudes were held in check by the humiliation of the Second World War, particularly with regard to the holocaust and Europe's need to be rescued by the US, Canada and Australia, all relatively Christian based nations at that point in their history. The evolutionary demons that went underground and disappeared from the radar screen for awhile, have now returned sevenfold. Since the end of the Second World War, not just Germany but the whole of Europe and, indeed, the western world has become increasingly secularized through the spreading evolutionary poison. Today the European Union is united in its secular vision of a new utopia, a new world order which already involves the abortion of unwanted children, the euthanasia of the feeble and those who are a strain on the welfare state, and a rising tide of anti-Semitism. The list of the weak and the hated will increase as these secularized European governments deal with their crumbling economies and social disintegration. It was a crumbling economy in Germany in the 1920's that lifted Hitler to the dais.

As has been written: "those who don't learn from history are doomed to repeat it", and "everything that goes round comes round". While many live in denial today believing the past record of the deadly consequences of evolutionary thinking on nations was just a historical aberration, the reality is that the toxic fruits of evolution are spreading everywhere. In an American Spectator article entitled, "Darwin and the Nazis", Dr. Weikart exposes the inroads of evolutionary thinking on contemporary life.

> BUT WHAT DOES THIS have to do with the present? Darwinists today are not Nazis.
> If you look back at the six points outlined above, however, you will find that many Darwinists today are advancing the same or similar ideas. Many leading Darwinists today teach that morality is nothing but a natural product of evolution, thus undermining human rights. E. O. Wilson, one of the most prominent Darwinian biologists in the world, and Michael Ruse, a leading philosopher of science (the latter is in Expelled) famously stated that ethics is "an illusion fobbed off on us by our genes."
> Many leading Darwinists today also claim that Darwinism undermines the Judeo-Christian concep-

tion of the sanctity of human life. Dawkins wrote in 2001 that we should try to genetically engineer an evolutionary ancestor to the human species to demolish the "speciesist" illusion that humans are special or sacred. In the same article he expressed support for involuntary euthanasia. Another critic of "speciesism," Peter Singer, one of the leading bioethicists in the world, argues that Darwinism destroyed the Judeo-Christian sanctity-of-life ethic, so infanticide and euthanasia are permissible. James Watson, one of the world's most famous geneticists and a staunch Darwinist, has railed at the idea that humans are sacred and special.

Today's Darwinists are not Nazis and not all Darwinists agree with Dawkins, Wilson, Ruse, Singer, or Watson. However, some of the ideas being promoted today by prominent Darwinists in the name of Darwinism have an eerily similar ring to the ideologies that eroded respect for human life in the pre-Nazi era.[92]

"THERE IS A WAY THAT SEEMS RIGHT TO A MAN, BUT IN THE END IT LEADS TO DEATH." Proverbs 14:12

Nations that embraced evolutionism in the twentieth century adopted a worldview where life and death were meaningless processes in a meaningless universe. There is no need to detail the record of atrocities committed under the atheistic communistic evolutionary and revolutionary regimes of Russia and China. These nations embraced evolutionism and its fruit of atheism and the dialectical materialism of Marxism with self-righteous and lethal zeal. Apparently Joseph Stalin, one of the Nephilim, the powerful human leaders of the twentieth century, once said: "killing a million people is no different than mowing your grass." (The Nephilim in Genesis 6 were the tyrants who ruled the world in the days of Noah prior to the flood.) Whether or not Stalin actually said it, he lived it. The body count of the Russian communists during their seventy year reign of terror is in the sixty million range "give or take" a few million. That is almost a million a year- a very powerful blood-stained lawn mower. This figure doesn't include the number of unborn children that were aborted under Russian Communism during their seventy years.

From an evolutionary perspective, the life of any human being is of no more value than that of a cat, a dog, a fish, an ant, or a blade of grass, for the simple reason that apart from GOD nothing has any ultimate value. We are only masses of molecules headed for oblivion. Although there is no scientific proof of evolution in action, transforming one kind of creature into another, there is plenty of evidence of evolution in action, slaughtering millions upon millions of people. Tragically and lamentably all too many live in the shadows of this deadly mythology where they cannot see the SON.

SACRIFICING CHILDREN TO THE EVOLUTIONARY GODDESS

One of the fruits of most demonic mythologies is the demand for the blood of the innocent, particularly the little children. Many of the Old Testament nations surrounding the people of Israel practised some form of child sacrifice to their idols. They believed that they would receive material blessings as a result of their sacrifices. When the people of Israel turned their backs on the living GOD, they began to sacrifice their children to these same idols. The LORD warned them through His prophets. When they ignored the warnings He handed them over into the brutal hands of their enemies. It is all documented in the historical record of the Old Testament that will never pass away. The prophet Jeremiah wrote:

> *"Hear the word of the LORD, O kings of Judah and people of Jerusalem. This is what the LORD Almighty, the God of Israel, says: Listen! I am going to bring a disaster on this place that will make the ears of everyone who hears of it tingle. For they have forsaken me and made this a place of foreign gods; they have burned sacrifices in it to gods that neither they nor their fathers nor the kings of Judah ever knew, and they have filled this place with the blood of the innocent. They have built the high places of Baal to burn their sons in the fire as offerings to Baal—something I did not command or mention, nor did it enter my mind. So beware, the days are coming, declares the LORD, when people will no longer call this place Topheth or the Valley of Ben Hinnom, but the Valley of Slaughter. In this place I will ruin the plans of Judah and Jerusalem. I will make them fall by the sword before their enemies, at the hands of those who seek their lives, and I will give their carcasses as food to the birds of the air and the beasts of the earth." (Jeremiah 19:3-7)*

From the time of their revolutions, the evolution-inspired regimes of Russia and China aborted unborn children as a sacrament or child sacrifice to their god of the atheistic state. As the evolutionary myth was absorbed into other nations, abortion followed in its path. Wherever evolution spread its poison, abortion was sure to follow. In our world today there are an estimated fifty to fifty-five million unborn children being sacrificed to the goddess of evolution every year. Canada sacrifices about one hundred thousand unborn children every year to Mother Earth, and the United States sacrifices over a million. We have inherited Stalin's bloody lawn mower, and used it on the helpless unborn.

The words of the LORD through Jeremiah are applicable to Canada and the United States today. I have lived and ministered in both countries. Both nations were formed and developed on some basic Judeo-Christian Biblical principles, specifically the reality of GOD as our Creator and Savior and the coming of a day of accountability and judgment. The rule of law in both nations was solidly anchored in the precepts of the Ten Commandments, which include a very definite and strong affirmation of the sanctity of human life. However, over the past century, as our nations have lived in the shadow of the evolutionary tree and begun to eat its fruit, the hearts and minds of many people have been drawn away from GOD to the basic elements of the world and the desires for the pleasures and treasures of the flesh. Having turned our faith away from GOD and His Holy Word and having trusted in the wisdom of humanity, our nations have resorted to the sacrifice of unborn children to solve the problems and receive the material blessings of this world that the evolutionary goddess has promised.

Once upon a time, in the land beneath our feet, almost everyone believed that an unborn child was a human being created in the very image of GOD with the right to life, liberty, and the pursuit of happiness. Even a Planned Parenthood pamphlet from 1963 spoke against abortion stating: *"An abortion requires an operation. It kills the life of a baby after is has begun."*[93]

The attitude toward the unborn changed when evolution ascended the stage and devalued their life, by calling them nothing but "blobs of tissue" that could be removed like cancerous growths. Medical doctors, lawyers, educators, politicians and even many of the clergy all bowed to this new deadly perspective on the life of the unborn child.

Yes, from an evolutionary viewpoint an unborn child is just a blob of tissue. This is also true for you and me, once the image of GOD is erased from the picture. If it is believed that we are not created in the image of GOD, then our lives have no intrinsic value and can be disposed of or incinerated with the garbage at the whim of those with power.

MORE TO COME

We do not have the time to tell of all of the poisonous fruit that the evolutionary tree produces for individuals and nations that live in its shadow. For the United States and Canada, we are only into the third generation that has been fed and watered at the evolutionary trough. Many of the fruits that are already present such as euthanasia, racism, eugenics, sexual promiscuity, the destruction of the institution of the family, and the disintegration of basic freedoms will continue to mature and become even more rotten. There is much more to come.

The events of 9-11 ought to have called us to sackcloth, ashes, and weeping before the LORD in repentance of our sin. Instead, it resulted in a self-righteous desire for revenge on our enemies. Like the people of Israel who were given over to the ruthless Babylonians, so our nations will be handed over to our enemies both within and without who hate and despise us. Without repentance in word and action, the fate of our nations is sealed. When our nations crumble and collapse, know that they have crumbled from the poison within and are no longer worthy of GOD'S protection from our enemies without. We will blame others as conditions deteriorate, but the mirror on the wall portrays the tragic truth.

Pray for the generation growing up today, for they are the ones who will experience a greater impact from the deadly evolutionary fruits than those of us who have had the privilege of growing up while our nations still had some remembrance of GOD and His Word. May many of the younger generations find their peace and hope in CHRIST.

THE TREE OF LIFE AND THE FRUIT OF THE SPIRIT

"The light shines in the darkness, and the darkness has not overcome it." John 1:5

If it weren't for CHRIST there would be no hope for our darkened and dying world. Praise GOD that there is another tree, the tree of Life and love from which GOD, by His HOLY SPIRIT, invites all of us to eat and drink through His SON JESUS. No one has to live in the shadows of evolutionary despair and pointlessness. JESUS said: **"The thief comes only to steal and kill and destroy; I have come that they may have life and have it abundantly." (John 10:10)** JESUS is the Way and the Truth and the Life: the abundant and eternal life, which GOD gives to all who believe.

There are a lot of reasons for concern and penitent persistent prayer for ourselves, families, neighbours, communities, nations and planet. The darkness is deep, but there is no reason for anyone to live a hopeless life, staring at the abyss of death. The LORD GOD our Creator is also the LORD GOD our Saviour who is merciful and forgiving to all who humble themselves, seek His face, and turn from their wicked ways to live in the love and truth of CHRIST.

In place of the fruits of evolution, we can live with the amazing fruit that the HOLY SPIRIT of GOD will produce in our lives as we eat and drink of the nutrients of GOD'S love and truth. It is written: **"the fruit of the Spirit is love, joy, peace, patience, kindness, goodness, faithfulness, gentleness and self-control." (Galatians 5:22,23)** We can experience this fruit of GOD'S SPIRIT in our lives and we can help others to come into the light of GOD'S presence and receive all of the love and truth they need to enjoy this fruit as well. May the LORD our GOD have mercy on us and help us to witness to and minister His grace and truth to those living in the land of the shadow of death.

"The night is almost over, the day is almost here." (Romans 13:12)

"On each side of the river stood the Tree of Life, bearing twelve crops of fruits, yielding its fruit every month. And the leaves of the tree are for the healing of the nations. No longer will there be any curse." (Revelation 22:2-3)

CONCLUSION

MULTITUDES IN THE VALLEY OF DECISION

"*Multitudes, multitudes in the valley of decision! For the day of the LORD is near in the valley of decision.*" (Joel 3:14)

HE'S EVERYTHING TO ME

"*In the stars His handiwork I see,
On the wind He speaks with majesty,
Though He ruleth over land and sea,
What is that to me?
I will celebrate nativity, for it has a place in history;
Sure He came to set His people free,
But, what is that to me?
Till by faith I met Him face to face,
And I felt the wonder of His grace.
Then I knew that He was more than just a GOD who didn't care,
Who lived a way out there,
And, now He walks beside me day by day,
Ever watching o'er me lest I stray.*

*Helping me to find that narrow way.
He is everything to me!*

Ralph Carmichael

KAIROS TIME FOR CHOOSING GOD

I believe that we are living in kairos time. The ancient Greeks had two main words for time: kronos and kairos. Kronos time was the passing of linear or ordinary time. Today kronos would be the counting of the seconds, minutes, hours and days of life. Kairos time, on the other hand, is a reference to a special or right time or opportune moment that will be significant or have an impact on an individual's life and world. I believe that today we are living in a supreme kairos time for choosing to trust in GOD. As GOD'S SPIRIT says: **"I tell you, now is the time of God's favor, now is the day of salvation."(2 Corinthians 6:2)**

Today is a day of amazing grace and salvation for many.

We are living in phenomenal times where the rapid expansion in knowledge and technological achievement is combined with an increase of evil and suffering, and the worldwide proclamation of the grace and truth of JESUS CHRIST. As in JESUS' parable of the weeds, good and evil are flourishing side-by-side (Matthew 13:24-30,36-43). This is the place, the day and the time where thousands and sometimes millions of people from every nation are turning their hearts back to GOD.

In Psalm 22 following a very vivid prophecy of CHRIST'S suffering and death on the cross and His resurrection victory, we are informed that a time will come when people from all parts of the world will turn to the LORD for His salvation. It is written:

> *All the ends of the earth*
> *will remember and turn to the LORD,*
> *and all the families of the nations*
> *will bow down before Him,*
> *for dominion belongs to the LORD*
> *and He rules over the nations...*
> *Posterity will serve Him;*
> *future generations will be told about the Lord.*
> *They will proclaim His righteousness,*
> *declaring to a people yet unborn:*
> *He has done it!*
> *(Psalm 22:27-31)*

"This is the day that the LORD has made and we will rejoice and be glad in it!"

There are multitudes in the valley of decision today. Decisions aren't always easy, but they are necessary and cannot be procrastinated forever. The most critical choice each individual has to make is the choice of what to believe regarding the ultimate realities of existence. What or who do I trust in and what is going to be the foundation, center and head of my life? There is only One who is worthy of our complete trust and deserves to be the LORD of each of our lives: the living GOD. Why? He is our Creator, the One who gave us life. He is our Provider, our Jehovah Jireh, who has provided all of the blessings of this life. He is our Savior, the only One Who gave His life for us and who can give eternal life. He is our Counselor and Helper and Comforter during our brief trip on this planet.

GOD in His wisdom and mercy has chosen to reveal Himself to our world. He has revealed His infinite and glorious nature through every detail of the material universe He created with all of its complexity and design. People often say that "seeing is believing". Although we can see the incredible world and the amazing universe, many choose not to recognize the awesome power and wisdom that brought it into existence and sustains it. In an even clearer way GOD has revealed Himself through His spoken and written Word. Through the prophets and apostles He has miraculously preserved a written historical testimony to His work in history and thereby provided us with a true and accurate communication of the truths concerning Himself, us, and our world. This testimony has been and is being preserved and distributed to all peoples for their understanding. Lastly, GOD Himself came in human form, in the person of His SON JESUS, who did an array of miraculous acts. CHRIST taught us about Himself and His will and showed us how to live an abundant and eternal life of love, joy, peace, and hope. **"Now, this is eternal life: that they may know You, the only true GOD, and JESUS CHRIST, whom You have sent!" JESUS, (John 17:3)** Our fallen race rejected JESUS and His message, and nailed Him to a cross. But that was the beginning and not the end of our salvation. CHRIST arose from the dead to fulfill all of His promises.

In the story JESUS told of the rich man and Lazarus, Abraham said these words: **"If they do not listen to Moses and the Prophets, they will not be convinced even if someone rises from the dead." (Luke 16:31)** When people refuse to listen to the truths that GOD has clearly revealed in His Word, then no matter what GOD does or has done, including the creation of an incredible universe and of life, or even the death and resurrection of His SON, these things will never be enough to cause them to acknowledge and humble themselves before Him. JESUS said: **"If I had not come and spoken to them, they would not be guilty of sin. Now, however, they have no excuse for their sin. He who hates Me hates My FATHER as well. If I had not done among them what no one else did, they would not be guilty of sin. But now they have seen these miracles, and yet they have hated both Me and My FATHER." (John 15:22-24)**

"Multitudes are in the valley of decision."

GOD AND SCIENCE

GOD established the laws of nature which normally govern our life on earth. Nevertheless, He is sovereign over all of His creation and can, according to His will, override those laws we are aware of and do miracles at any time. GOD'S thoughts and ways are not contrary to science, just light years ahead of it. The miracles of Scripture involve not only physical realities but metaphysical ones. Theologians and others who seek to confine GOD and Christian faith to the limited scientific understanding of any age will never understand true Christianity. Their god is way too small.

Our world and GOD'S Word are full of miracles. Anyone who is a Christian has to have open eyes to see the miracles that surround us every day and believe in the miracles testified to in the Bible. Without miracles, the Christian faith is empty and hopeless, for then there would be no CHRIST. I have experienced a number of miracles in my own life and in the lives of people I know well. If you do not believe in the living GOD who does miracles then, of course, you have to explain every miracle including creation with a materialistic explanation of some kind. Either GOD is a miracle-working GOD and the Creator of the cosmos or He isn't GOD.

The proper jurisdiction of the true science is, through intelligently designed experiments and observations, to gain a greater understanding about our present universe and world which GOD has created for us. Trying to explain the origin of life in the distant past through material causes alone isn't legitimate science. The origin of our universe and life involves the creation of incredibly complex information and this isn't something that is repeatable. Since science deals with things that can be repeated and demonstrated, to use science to explain our origins is to cease using it as a methodology and to transform it into a godless materialistic philosophy which rejects GOD'S intelligent design and replaces it with the mythology of evolution. Is our universe and world the result of a mindless fluke of materialism or the creation of the infinite mind of the incredibly awesome GOD?

"Multitudes are in the valley of decision."

THE FENCE SITTERS

There are certainly many who are unwilling to confront the logical conclusions of evolution and at the same time are unwilling to give up their belief in a god and some kind of eternal paradise. Theistic evolutionists are the proverbial fence sitters. They precariously position themselves on the edge of faith, but sooner or later they have to put their foot down either in Darwin's materialistic land where the atom reigns or in CHRIST'S Kingdom where His Word is truth. Either CHRIST is your GOD and your Master and His HOLY SPIRIT and infallible Word is your guide, or science is your god and master and its infallible word is your guide. Science is not the source or salvation of our lives. We cannot have two Kings or LORD'S at the center of our life.

> *"The GOD Who made the world and everything in it is the LORD of Heaven and Earth...He Himself gives all men life and breath and everything else... GOD did this so that men would seek Him and perhaps reach out for Him and find Him, though He is not far from each one of us. For in Him we move and live and have our very being" (Acts 17:24-28)*

> *"Multitudes are in the valley of decision."*

DIFFERENT UNIVERSES

I realize and understand, as many websites and blogs reveal, that from the secular materialistic perspective, Christians are seen as being unintelligent, ignorant people who can't face up to or cope with the materialistic realities of life. We are accused of living in denial of reality and in superstitious belief systems. You see, we may all live on the same planet Earth but we live in very different universes. Evolutionists live in a purposeless universe that just happened out of nothing and is returning to nothing, and in which incredibly complex material realities appear out of nowhere and dissipate into nowhere. Nevertheless, they zealously believe that their own accidental intelligence combined with the philosophy of scientific materialism is an inerrant guide to understanding the futile universe they have imagined.

If I didn't believe in or know CHRIST it would be easy to believe in evolution, not because it's proven, but because it is the only materialistic explanation for life and without GOD that is all there is. However, when people, through the ministry of the HOLY SPIRIT and the preaching of GOD'S Word, repent of their sin and come to believe in and know JESUS CHRIST as their SAVIOR and LORD, they are born again. They begin to see everything in GOD'S creation from a completely different perspective: the eternal perspective of GOD and His Word, and not the temporal materialistic perspective of humanity. With the eyes of faith we see an incredibly marvellous world and universe created for us by our omnipotent and gracious FATHER and LORD. We see an amazing world which is quickly deteriorating through humanity's evil, but which will be replaced with an even grander eternal paradise for those who love and trust in GOD and His Word of truth. What do you see?

"Multitudes are in the valley of decision."

GOD IS

GOD is the eternal Creator who planned and formed our world and each of us. He is the great Engineer of the universe, the great Biologist and Chemist of the genetic code, the great Artist who has endowed all of creation with beauty, and the great Author of His Word and of our salvation. He is all of this and much more. His wisdom is beyond our understanding and His ways beyond our grasp. GOD is perfectly just, righteous, holy, perfect, loving, kind, patient, and forgiving. He always has been and always will be.

It is written:

"Now fear the LORD and serve Him with all faithfulness...But if serving the LORD seems undesirable to you, then choose for yourselves this day whom you will serve...But as for me and my house, we will serve the LORD. (Joshua 24:14-16)

We are living in the most exciting and challenging days for believers and for the Christian church. It is time to choose to follow our GOD, our Creator, Savior and Counselor.

GOD'S Kingdom, the Kingdom of JESUS CHRIST, is soon going to come- *"the zeal of the LORD Almighty will accomplish this" (Isaiah 9:7)* As someone else has written, soon the Director will step onto the stage and the world as we know it will be over. The world to come is more magnificent, more beautiful than anything we have yet experienced. For those who love CHRIST and are waiting for Him GOD has an amazing new world to show us. As it is written: *"No eye has seen, no ear has heard, no mind has conceived what God has prepared for those who love him" (1 Corinthians 2:9)*

Is your hope in this world which will soon pass away or in GOD'S new world which will soon be here? What will you decide? Whom will you choose to serve? The living Triune GOD of the Bible: FATHER, SON and HOLY SPIRIT, or the Little Tinkerer of evolutionary mythology?

"Multitudes are in the valley of decision."

THIS IS MY GOD

I conclude with a collection of Scripture verses put together by our daughter Heidi during the time she was diagnosed with and went through surgery and treatment for thyroid cancer. She was fifteen at the time. She is doing well today, thanks to the LORD our GOD, her Creator, and the very gifted medical doctors and nurses who used their understanding and skills to help heal her. Her writing is titled: *"This Is My GOD"*. We hope and pray that each of you who are reading this book will know Him as your GOD as well, because He is your GOD- your Creator, Savior and Counsellor, who knows and loves you and died to give you eternal life. GOD IS who He is and is worthy of all of our trust and praise.

THIS IS MY GOD
By Heidi Pinno

GOD, who answered me in the day of my distress
and who has been with me wherever I have gone
(Genesis 35:3b)
GOD, who is greater than all other gods (Exodus 18:11)
GOD, who broke the bars of my yoke
and enabled me to walk with my head held high
(Leviticus 26:13b)
GOD, who is not a man that He should lie (Numbers 23:19)

GOD IS!

GOD, whose works are perfect and whose ways are all just (Deuteronomy 32:4)
GOD, who fights for me, just as He promised (Joshua 23:10)
GOD, whose presence causes even the mountains to quake (Judges 5:5)
GOD, who gives offspring (Ruth 4:12)
GOD, who does not look at the things man looks at, but looks at the heart (1 Samuel 16:7)
GOD, whose word is trustworthy (2 Samuel 7:28)
GOD, who cannot be contained even in the highest heavens (1 Kings 8:27)
GOD, who hears my prayer and sees my tears (2 Kings 20:5)
GOD, whose dwelling place is full of strength and joy (1 Chronicles 16:27)
GOD, who is just (2 Chronicles 12:6)
GOD, whose love endures forever (Ezra 3:11)
GOD, who is worshipped by the multitudes of heaven (Nehemiah 9:6)
GOD, who wounds but also binds up, who injures but also heals (Job 5:18)
God, who is the stronghold of my life (Psalm 27:1)
GOD, who is my confidence (Proverbs 3:26)
GOD, who will bring every deed into judgment, whether good or evil (Ecclesiastes 12:14)
GOD, who calls me beautiful, oh so beautiful (Song of Songs 4:1)
GOD, who is wonderful in counsel and magnificent in wisdom (Isaiah 28:29)
GOD, who is my refuge in times of distress (Jeremiah 16:19)
GOD, whose great love keeps me from being consumed (Lamentations 3:22)
GOD, who does NOT take pleasure in the death of anyone (Ezekiel 18:32)
GOD, who is the great and awesome GOD (Daniel 9:4)

GOD, who loves to redeem all people (Hosea 7:13)
GOD, who is slow to anger and abounding in love
(Joel 2:13)
GOD, who reveals His thoughts to men (Amos 4:13)
GOD, whose day of judgment is coming soon for all nations
(Obadiah 15)
GOD, who brought my life up from the pit (Jonah 2:6)
GOD, who delights in showing mercy (Micah 7:18)
GOD, who has clouds as the dust beneath His feet
(Nahum 1:3)
GOD, whose eyes are too pure to look upon evil
(Habakkuk 1:13)
GOD, who quiets me with His love (Zephaniah 3:17)
GOD, who is with me (Haggai 1:13)
GOD, who will save His people from the countries of the
east and west (Zechariah 10:1)
GOD, who does not change (Malachi 3:6)
GOD, who is gentle and humble in heart (Matthew 11: 29)
GOD, who with nothing is impossible (Mark 10:27)
GOD, whose name is holy (Luke 11:2)
GOD, who has overcome the world (John 16:33)
GOD, who does not show favoritism (Acts 10:34)
GOD, who will give to each person according to what he
has done (Romans 2:6)
GOD, who provides a way out of every temptation
(1 Corinthians 10:13)
GOD, who comforts us in all our troubles (2 Corinthians 1:4)
GOD, who does not judge by external appearance
(Galatians 2:6)
GOD, who is rich in mercy (Ephesians 2:4)
GOD, who began a good work in me and will carry it on to
completion (Philippians 1:6)
GOD, who rescued me from the dominion of darkness and
qualified me to share in the inheritance of the saints in the
kingdom of light (Colossians 1: 12)

GOD, who tests our hearts (1 Thessalonians 2: 4)
GOD, who will strengthen and protect me from the evil one
(2 Thessalonians 3:3)
GOD, Who has unlimited patience (1 Timothy 1:16)
GOD, who did not give me a spirit of timidity,
but a spirit of power, of love, and of a sound mind
(2 Timothy 1:7)
GOD, who does not lie (Titus 1:2)
GOD, in whom we have many good things (Philemon 6)
GOD, who tasted my death (Hebrews 2:9)
GOD who can sympathize with my weaknesses,
and who has been tempted in every way, yet was
without sin (Hebrews 4: 15)
GOD, who promises to come near to me if I draw near to
Him (James 4:8)
GOD, who has given me new birth into a living hope
(1 Peter 1:3)
GOD, who is not slow in keeping His promises (2 Peter 3:9)
GOD, who is light, in Him there is no darkness (1 John 1:5)
GOD, whose command is to walk in love (2 John 6)
GOD, whose mercy brings us into eternal life (Jude 21)
GOD, who is alive forever and ever (Revelation 1:18)

GOD Is Who He Is! Hallelujah!

"Amen! Come LORD JESUS!
The grace of the LORD JESUS be with
GOD'S people. Amen!"
Revelation 22:20,21.

GOOD NEWS FROM OUR CREATOR

1. GOD created us in His own image to be His children. He loves us and desires that we should love and obey Him, and love and serve others. (Matthew 22)

2. All of us have sinned against GOD, others, and ourselves. Through sin we are destroying everything and heading towards death and eternal judgment. (Romans 3; Revelation 20)

3. GOD, in His love and amazing grace, sent His SON, JESUS CHRIST, into our world. JESUS gave His life on a cross for all of our sin. He rose again from the dead to give eternal life to everyone who sincerely believes in Him. (John 3; Rom. 3)

4. GOD the FATHER, SON and HOLY SPIRIT, calls upon each of us to repent of our sin and trust in Him. JESUS preached: "Repent for the Kingdom of Heaven is near!" (Matthew 4)

5. CHRIST will soon return to judge our planet and establish His new eternal world for His people. (Matthew 24-25; Revelation 21)

"For GOD so loved the world that He gave His one and only Son, that whoever believes in Him will not perish but have eternal life."
JESUS (John 3:16)

If you believe this, you need to ask GOD to forgive your sins, and place your faith in CHRIST and what He has done for you. GOD can and will forgive all of your sins and give you the gift of eternal life. JESUS promises you that!

May it be so! Come LORD JESUS!

ENDNOTES

1. www.brainz.org/15-coolest-cases-biomimicry
2. www.creation.com/the-universe-is-finely-tuned-for-life
3. Grant Jeffrey, *Creation:Remarkable Evidence of GOD'S Design,* Frontier Research Publications, Toronto, ON 2003 p.117,118
4. Henry M. Morris, Editor, *Scientific Creationism,* Creation –Life Publishers, San Diego, CA, 1974, p. 32
5. www.en.wikipedia.org/wiki/Fred_Hoyle, "The Universe: Past and Present Reflections", Engineering and Science, November 1981, pp. 8-12
6. www.godandscience.org/apologetics/quotes.html Ellis, G.F.R. 1993. The Anthropic Principle: Laws and Environments. *The Anthropic Principle*, F. Bertola and U.Curi, ed. New York, Cambridge University Press, p. 30.
7. www.godandscience.org/apologetics/quotes.html John O'Keefe quoted in Heeren, F. 1995. Show Me God. Wheeling, IL, Searchlight Publications, p. 200
8. www.godandscience.org/apologetics/quotes.html Arthur Eddington quoted in Heeren, F. 1995. *Show Me God*. Wheeling, IL, Searchlight Publications, p. 233

9. www.godandscience.org/apologetics/quotes.html Jastrow, R. 1978. *God and the Astronomers.* New York, W.W. Norton, p. 116.
10. www.godandscience.org/apologetics/quotes.html Wernher von Braun quoted in McIver, T. 1986. *Ancient Tales and Space-Age Myths of Creationist Evangelism.* The Skeptical Inquirer 10:258-276.
11. www.brainyquote.com/quotes/authors/w/wernher_von_braun.html
12. www.simpletoremember.com/articles/a/creatorfacts Michael Turner quoted by Dr. Gerald Schroeder, *"The Fine Tuning of the Universe"*
13. Michael Denton, *Evolution: A Theory in Crisis* (Chevy Chase, MD: Adler and Adler Publishers, Inc. 1986) pgs 328,342
14. Werner Gitt, "Creation ex nihilo" Vol. 20 No.1,December-February 1997, Answers in Genesis
15. www.creation.com/journal-of-creation-213 Alex Williams "Astonishing DNA complexity demolishes neo-Darwinism" 114 JOURNAL OF CREATION 21(3) 2007
16. www.creation.com/journal-of-creation-213 Alex Williams "Astonishing DNA complexity demolishes neo-Darwinism" 114 JOURNAL OF CREATION 21(3) 2007
17. "Born Again" www.uncommon-descent.com/intelligent-design/it-doesnt-matter-what-we-name-them/#comments
18. www.thethinkingbusiness.co.uk/brain_facts.html Tony Buzan, Head Strong 2001)
19. www.lifeoptimizer.org/2008/05/08/top-10-interesting-facts-about-the-brain/
20. www.talkorigins.org/faqs/origin/chapter6.html Charles Darwin, *"On the Origin of Species by Means*

of Natural Selection, or the Preservation of Favoured Races in the Struggle for Life," 1859 Chapter 6
21. www.webcache.googleusercontent.com/search?q=cache:http://www.atschool.edu web.co.uk/sbs777/vital/evolution html H.S.Hamilton (MD) *"The Retina of the Eye - An Evolutionary Road Block"*
22. Jobe Martin *The Evolution of a Creationist* Biblical Discipleship Publishers, Rockwall, Texas 2004
23. Robert E. Kofahl & Kelly L. Seagraves *The Creation Explanation*, Harold Shaw Publishers, Wheaton, Ill 1975 pgs. 6,7
24. Geoff Simmons, *Billions of Missing Links*, Harvest House Publishers, Eugene, Oregon, 2007, p. 22
25. Michael Denton, *Evolution: A Theory in Crisis* (Chevy Chase, MD: Adler and Adler Publishers, Inc. 1986) pg 358
26. creation.com/muggeridge-eldredge Malcolm Muggeridge, "The End of Christendom: Inaugural address of The Pascal Lectures on Christianity and the University,"Ontario Canada, University of Waterloo,1978
27. creation.com/amazing-admission-lewontin-quote Richard Lewontin, Billions and billions of demons, The New York Review, p. 31, 9 January 1997.
28. www.creationsafaris.com/crev200710.htm
29. Henry M. Morris, *The Genesis Record* Baker book House, Grand Rapids, Michigan, 1976, pgs 17,18
30. Henry M. Morris, Ibid pgs. 19,20
31. Phillip E Johnson, *Defeating Darwinism by Opening Minds*, InterVarsity Press, Downers Grove ILL 1997, p. 72,73
32. Werner Gitt, *In the Beginning Was Information*, CLV Christliche Literatur-Verbreitung, Bielefeld, Germany, 1997 p. 47

33. www.en.wikipedia.org/wiki/Fred_Hoyle
34. Gary Parker, *Creation: Facts of Life*, Master Books, Green Forest, AR, 2008 edition, p. 11
35. Gary Parker Ibid p. 29
36. www.thegrandexperiment.com
37. www.en.wikipedia.org/wiki/Dean_H._Kenyon
38. www.creationsafaris.com/crev200706.htm#20070630a
39. www.whyevolution.com/chimps.html
40. www.ornl.gov/sci/techresources/Human_Genome/faq/compgen.shtml
41. www.darwinismrefuted.com/myht_of_homology_04.html
42. Skeptics Vs. Creationists: A Formal Debate Hosted by the Syndney Morning Herald, Creation Ministries International, Eight Mile Plains, Australia, 2006, p. 37
43. Charles Darwin, *"On the Origin of Species by Means of Natural Selection, or The Preservation of Favoured Races in the Struggle for Life"* p. 184
44. Carl Werner, *Evolution: The Grand Experiment*, New Leaf Publishing Group, Green Forest, AR, 2007, p. 54
45. John C. Sanford, *Genetic Entropy & the Mystery of the Genome*. Technical Publisher: Elim Publishing 2005.
46. www.en.wikipedia.org/wiki/John_C._Sanford
47. Jonathon Sarfati, *The Greatest Hoax on Earth?* Creation Book Publishers, Atlanta, GA 2010
48. Carl Werner, *Evolution: The Grand Experiment,* New Leaf Publishing Group, Green Forest, AR, 2007, p. 85
49. Carl Wieland, *Stones and Bones: Powerful Evidence Against Evolution*, Master Books, Green Forest, AR, 2005, p. 11-12
50. www.cbsnews.com/stories/2007/10/15/tech/main3368505.shtml

51. Charles Darwin, *The Origin of Species By Means of Natural Selection or the Preservation of the Favoured Races in the Struggle for Life,* 1st edition, John Murray, London, 1859 www.talkorigins.org/faqs/origin.html
52. Carl Werner, *Evolution: The Grand Experiment,* New Leaf Publishing Group, Green Forest, AR, 2007, p. 74
53. en.wikipedia.org/wiki/Mary_Higby_Schweitzer
54. www.americanhumanist.org/who_we_are/about_humanism/Humanist_Manifesto_I
55. www.americanhumanist.org/who_we_are/about_humanism/Humanist_Manifesto_II
56. www.american humanist.org/who_We_Are/About_Humanism/Humanist_Manifesto _III
57. www.thebereancall.org/node/2348
58. www.creationsafaris.com/wgcs_toc.htmc. 2000 David F. Coppedge, Master Plan Productions
59. www.creation.com/why-most-scientists-believe-the-world-is-old
60. Jobe Martin, *The Evolution of a Creationist,* Biblical Discipleship Publishers, Rockwell, Texas, p.44
61. Jobe Martin Ibid p. 44 quoting from G.A Kerkut, Implications of Evolution New York: Pergamon Press, 1960, p. 6
62. John K. G. Kramer chapter 3, *In Six Days: why fifty scientists choose to believe in creation,* edited work John Ashton, Master Books, Green Forest, AR 2003, p. 47
63. Jonathon Sarfati, *The Greatest Hoax on Earth?* Creation Book Publishers, Atlanta, GA 2010 pgs. 183-199
64. Jonathon Sarfati, Ibid. pgs. 200-222
65. Danny Faulkner, *Universe by Design: An Explanation of Cosmology and Creation,* Master Books, Green Forest, AR, 2004, pgs.101-104

66. Danny Faulkner, Ibid. pgs.101-104
67. Phillip E Johnson, *Darwin on Trial* InterVarsity Press, Downers Grove, ILL 2nd Edition, 1993, p. 28
68. Eugenie C. Scott, *Evolution Vs. Creationism,* University of California Press Berkeley and Los Angeles,2005, p. 3
69. www.en.wikipedia.org/wiki/Albert_Einstein#World_ War_II_and_the_Manhattan_ Project
70. Barna Poll Shows Sad State of Christians www.fcov. blogspot.com/2009/03/well-new-barna-poll-came-out-today-and.html
71. Paul A. Zimmerman, *Creation, Evolution and GOD'S Word,* Concordia Publishing House, St. Louis, MI, 1972, pgs 120-122 73.
72. www.phrases.org.uk/meanings/red-in-tooth-and-claw.html
73. C. S. Lewis, *"The Problem of Pain"* quoted in www.handsupport.org/newsletters /winter99.pdf
74. Ravi Zacharias and Norman Geisler, *Who Made GOD*, Zondervan, Grand Rapids, Michigan, 2003, p.30
75. Jonathan Sarfati, *Refuting Compromise* Master Books, Green Forest, AR, 2004, pgs 67-106
76. Martin Luther, "A Short Confession on the Holy Sacrament"
77. Frank Zindler, American atheist, in a debate with William Craig, Atheism vs Christianity video, Zondervan, 1996 Evolution and Atheism http://bevets.com/evolution.htm
78. 'I don't believe Jesus was God' -United Church's new moderator rejects Bible as history book- Bob Harvey "The Ottawa Citizen" Friday 24 October 997 http:// http://webcache.googleusercontent.com/search?q=cache:http://www.igs.net/~tonyc/mod1.html

79. Nature, Vol. 394, No. 6691, p. 313 (1998) Macmillan Publishers Ltd.
80. "From Atheism; and the Scientific Community" www.nwcreation.net/atheism.html
81. Jonathon Sarfati, " National Academy of Science is godless to the core — Nature Survey" www.creation.com/national-academy-of-science-is-godless-to-the-core-survey
82. Will Provine, "Evolution: Free will and punishment and meaning in life," www.fp.bio.utk.edu/darwin/1998/provine_abstract.html.
83. Will Provine, "Academe" January 1987 pp.51-52 †
84. Will Provine, *Darwinism: Science or Naturalistic Philosophy* April 30 1994 Quoted on Stephen E. Jones Website www.bevets.com/equotesp5.htm
85. Bertrand Russel, *Mysticism and Logic* (1981) p.41 Quoted fy Stephen E. Jones Website www.bevets.com/equotesr.htm#R
86. Richard Dawkins, *The Blind Watchmaker*, New York: Norton. p. 6 quoted at www.wikipedia.org/wiki/Richard_Dawkins
87. Richard Dawkins, "God's Utility Function," published in Scientific American (November, 1995), p. 85
88. Gil Dodgen, www.uncommondescent.com/education/does-atheism-poison-everything-debate-between-david-berlinski-and-christopher-hitchens/#comments Uncommon Descent, September 7, 2010
89. R.J Rummel, "Death by Government" www.hawaii.edu/powerkills/NOTE1.HTM
90. Richard Weikart "From Darwin to Hitler: Evolutionary Ethics, Eugenics, and Racism in Germany", www.spectator.org/archives/2008/04/16/darwin-and-the-nazis

91. Denyse O'Leary, *By Design or by Chance*, Castle Quay Books Kitchener, ON, 2004 p.72
92. Richard Weikart, From Darwin to Hitler: Evolutionary Ethics, Eugenics, and Racism in Germany (Palgrave Macmillan). http://spectator.org/archives/2008/04/16/darwin-and-the-nazis
93. www.freerepublic.com/focus/f-news/1185302/posts

BIBLIOGRAPHY

1. Ashton, John, editor, *In Six Days: why fifty scientists choose to believe in creation,* Green Forest, AR: Master Books, 2003.

2. Darwin, Charles, *On The Origin of Species by Means of Natural Selection, or The Preservation of Favored Races in the Struggle for Life,* London: 1st edition, John Murray, 1859.

3. Denton, Michael, *Evolution: A Theory in Crisis,* Chevy Chase, MD: Adler and Adler Publishers, Inc. 1986.

4. Faulkner, Danny, *Universe by Design: An Explanation of Cosmology and Creation*, Green Forest, AR: Master Books, 2004.

5. Gitt, Werner, *In the Beginning Was Information,* Bielefeld, Germany: CLV Christliche Literatur-Verbreitung, 1997.

6. Jeffrey, Grant, *Creation: Remarkable Evidence of GOD'S Design,* Toronto, ON: Frontier Research Publications, Inc., 2003.

7. Johnson, Phillip E., *Darwin on Trial,* Downers Grove, ILL: InterVarsity Press, 2nd Edition, 1993.

8. Johnson, Phillip E., *Defeating Darwinism by Opening Minds,* Downers Grove ILL: InterVarsity Press, 1997.

9. Kofahl, Robert E. & Seagraves, Kelly L., *The Creation Explanation,* Wheaton, Ill: Harold Shaw Publishers, 1975.

10. Martin, Jobe, *The Evolution of a Creationist,* Rockwall, Texas: Biblical Discipleship Publishers, 2004.

11. Morris, Henry M. Editor, *Scientific Creationism*, San Diego, CA: Creation-Life Publishers, 1974.

12. Morris, Henry M., *The Genesis Record* Grand Rapids, Michigan: Baker book House, 1976.

13. O'Leary, Denyse, *By Design or by Chance,* Kitchener, ON: Castle Quay Books 2004.

14. Parker, Gary, "Creation: Facts of Life", Green Forest, AR:, Master Books, 2008

15. Sanford, John C., *Genetic Entropy & the Mystery of the Genome.* Technical Publisher: Ivan Press 2005

16. Sarfati, Jonathan, *Refuting Compromise*, Green Forest, AR: Master Books, 2004

17. Sarfati, Jonathan, *Refuting Evolution,* Green Forest, AR: Master Books, 2000

18. Sarfati, Jonathon, *The Greatest Hoax on Earth?* Atlanta, GA: Creation Book Publishers, 2010

19. Scott, Eugenie C., *Evolution Vs. Creationism,* Los Angeles, CA: University of California Press Berkeley and 2005

20. Simmons, Geofrey, *Billions of Missing Links*, Eugene, Oregon: Harvest House Publishers, 2007

21. Werner, Carl, *Evolution: The Grand Experiment,* Green Forest, AR: New Leaf Publishing Group, 2007

22. Wieland, Carl, *Stones and Bones: Powerful Evidence Against Evolution*, Green Forest, AR: Master Books, 2005

23. Zacharias, Ravi and Geisler, Norman, *Who Made GOD*, Grand Rapids, Michigan: Zondervan, 2003

24. Zimmerman, Paul A Editor, *Creation, Evolution and GOD'S Word,* St. Louis, MI: Concordia Publishing House, 1972

25. Skeptics Vs. Creationists: A Formal Debate Hosted by the Syndney Morning Herald, Eight Mile Plains, Australia: Creation Ministries International, 2006

LaVergne, TN USA
02 March 2011
218427LV00001B/2/P